a VOICE OF HER OWN

Edited by
Thelma Poirier, Doris Bircham, JoAnn Jones-Hole,
Anne Slade, and Susan Ames Vogelaar

Published by the
University of Calgary Press
2500 University Drive NW
Calgary, Alberta, Canada T2N 1N4
www.uofcpress.com

We acknowledge the financial support of the Government of Canada, through the Book Publishing Industry Development Program (bpidp), and the Alberta Foundation for the Arts for our publishing activities. We acknowledge the support of the Canada Council for the Arts for our publishing program.

LIBRARY AND ARCHIVES CANADA CATALOGUING IN PUBLICATION

A voice of her own / edited by Thelma Poirier ... [et al.].

(Legacies shared ; 18)
ISBN 1-55238-180-3

1. Women ranchers – Alberta – Interviews. 2. Women ranchers – Saskatchewan – Interviews. I. Poirier, Thelma II. Series.

FC3238.V63 2005 971.23'0082
C2005-904772-0

Cover design, Mieka West.
Internal design & typesetting, zijn digital.

This book is printed on acid-free paper.

Printed and bound in Canada by AGMV Marquis

TO ALL THE RANCH WOMEN OF WESTERN CANADA

TABLE OF CONTENTS

ROCKY MOUNTAIN VIEW

WEST OF THE ROCKIES, KAMPLOOPS

SOUTHWEST ALBERTA FOOTHILLS

CYPRESS HILLS

WOOD MOUNTAIN UPLANDS

PRAIRIE TO PARKLANDS

ABOUT THE EDITORS

INTRODUCTION

Saddles, horses, cows, and grass. These are the things that differentiate the lives of ranch women from the lives of other women, urban or rural. Just as they underlined the lives of past generations of ranch women, they underline the everyday lives of contemporary ranch women, whether they are in the corral or the kitchen, or at a desk behind a computer. Without the land, the grass, the livestock, they would not be ranch women.

As a mother protects a child, a ranch woman protects the ranch, both the existence of the ranch and the state of the ranch. If this means rounding up yearlings or checking water, that is what she does. If it means going out and getting a job in town to help pay the bills, that is what she does. Or if it means political involvement, then she may be politically involved. In whatever she does, the ranch most always motivates her.

A ranch woman can view a ranch in many ways. It may be a business. The bottom line is usually related to the profit margin, but often those concerns are shared or sometimes they are left to a male partner. Over the past century, the roles of women have changed, and today's women are usually very involved in marketing and accounting.

Ranching is also a lifestyle. Sometimes the emphasis is on a traditional lifestyle, doing things the way they were done in the "old" West. Sometimes the emphasis is on modern techniques. For sure, there is no single way to manage a ranch. Whatever way or combination of ways is adopted, it is usually unique to the ranch in question. Sometimes there is a tendency to do things a certain way in one community and not in another. At the same time, there will always be those who deviate – those who are more traditional or more modern than most. Some round up only with horses, some use dogs, and others use an airplane.

To most ranch women, the ranch is more than just an industry or a lifestyle. It is a relationship with a place, with the land. It is the expanse of natural prairie, moss phlox that flowers in May, the needle-and-thread grass that blossoms in July, the antelope that graze among the cows, the ferruginous hawk overhead. Perhaps, because so much of the land on ranches remains in a natural state, ranch women have especially close bonds with the land. Perhaps, because each spring the spirit of the land is born again and because women's lives are so focussed on birth, the bond is even stronger. When Marjorie Linthicum wrote, "I have a special feeling for this land," in a brief presented to the hearings for the Grasslands National Park she spoke on behalf of all ranch women who have watched life play out its cycles through birth and growth, maturity and death in a natural environment. When human relationships, friendships, marriages, families are moulded in the context of the land and the livestock, the bonds are even stronger. Most people in the twenty-first century will only visit such places, go with friends or family to vacation in the natural spaces. Ranch women live there season in and season out.

If ranch women are pragmatic or tenacious, it is out of necessity. These are qualities that develop in a life that is defined by wind and drought, snow and ice. The land impacts upon the human spirit, mixing pragmatism with empathy, tenacity with tenderness.

In 1999 I was invited to organize a presentation by women for the Canadian Cowboy Festival in Calgary. It was to be in February, and the first challenge would be to leave the ranch, the daily chores behind. That was arranged. On then to selecting women to take part in the presentation. I knew some of the women who would be at the Festival: two poets from the Maple Creek area, Doris Bircham and Anne Slade; one from Pincher Creek, Susan Ames Vogelaar. I had read the stories of Isabella and Bobbi June Miller, two rodeo competitors from west of Calgary; and then I knew Jo Ann Jones-Hole, who had completed a history of the Calgary Bull Sale. A telephone call revealed that Judy Alsagar, formerly of the Gang Ranch in British Columbia, would be passing through Calgary at festival time. And Heather Harden, a young woman from my hometown, was working on the Wineglass Ranch west of Cochrane, and maybe her employer, Edith Wearmouth, the owner of the Wineglass, would take part. What about Elizabeth Ebert, the poet from South Dakota, and Liz Masterson, the singer from Colorado? I took along Jody Hordenchuk, a teenager from Wood Mountain, and finally I had a roster of twelve. The presentation was coming together, even though no one would be paid for being there and everyone had ranch chores to leave behind. Obviously this was important.

The panel featuring the ranch women started with a scattering of people in a large room. Before it ended, there was not even standing room. The stories were poignant and humorous, they were contemporary and they were true. That was the clincher, they were true. One by one, the twelve ranch women revealed the joys and sorrows of their lives in the mountains and on the plains. The audience heard how pigs paid for a ranch, how the Gang Ranch was "won and lost," how a young girl grew up in a bunkhouse, how another had to fill her father's boots at age twelve, how a son was buried on a ranch, how a senior went back to working the land when her husband's health failed. One woman told how the floor in her first ranch home dropped a foot from one wall to the other.

Dale Evans was signing autographs in the hallway while the ranch women told their life stories at the Convention Centre. Dale epitomized the "cowgirl" image of the West – the glamour, the style, the Hollywood mythology. The ranch women on the panel were hardly cowgirls in the same sense. From the stories they told, most of them did not think of themselves as cowgirls – their lives reflected more grit and less glamour. Later interviews would reveal that few ranch women associate themselves with the term *cowgirl*. The interviews underline the absence of the word. Unless they were referring to a childhood wish "to be a cowgirl" or to young women in rodeo competition, the word was seldom spoken. The reason may be that the girl in *cowgirl* does not reflect the maturity of ranch women's lives; it only seems to apply to the more recreational aspects of their lives, such as rodeo or cowboy poetry. Imagine the surprise if someone were to advertise a "cowgirl poetry" gathering. However, it must be noted that many books have been published under a "cowgirl" title, even books that are more about ranch women than the legends of arenas and stages. In 1998 I put together an anthology of writing by North American ranch women, and Red Deer College published it under the title *Cowgirls: One Hundred Years of Writing the Range*. While the term *cowgirls* may not have always described the lives of the women accurately, it catches the attention of many consumers.

The day after the panel discussion, an interviewer from CBC Radio asked me if I thought the stories of the women on the panel should be a book. That was when I realized they hadn't even been taped. Yet they were still there, they still existed, as did the stories of hundreds and thousands of ranch women. Teresa Jordan had interviewed several dozen ranch women in the western United States and published their stories under the title *Cowgirls*. In another book titled *Cowgirls*, Candace Savage gave credence to the first ranch women of North America, a few of whom were Canadian, along with the stars of the arena and screen. She extended the same credence to children in another book, *Born to Be a Cowgirl*. Other anthologies, such as *Wild West Women* by Rosemary Neering (1995), hint at the lives of Canadian ranch women.

In 1999 I could not find the parallel to Teresa Jordan's book in Canada, not one that allowed contemporary ranch women of Canada to tell their own stories, that allowed them a voice of their own. I knew the preparation of such a book would be a challenge. Among the thousands of women in agriculture, many resided on ranches, their chief source of income coming from the ranching industry. There wasn't going to be a book to include all of them, but maybe a representative, or even a random group of stories could be collected. Women of all ages, from various geographical areas and with differing educations could be approached.

Having just completed an anthology of writing by ranch women, I wasn't prepared to accept the challenge alone. Several of the women on the Calgary panel were writers, and the very first four I spoke to accepted the challenge. JoAnn Jones-Hole, Doris Bircham, Anne Slade, and Susan Ames Vogelaar agreed to work with me. We would each do some interviewing and we would plan the book, not sure how it would be published. Mobil Oil of Calgary provided some start-up funding. We talked to people at the University of Calgary Press and soon Janice Dickin came on board as a consultant. Our goal was to publish a book by 2005, to allow women who grew up and lived on ranches during the first one hundred years of Alberta and Saskatchewan to tell their stories. Three years later we realized that the more women we interviewed, the more we wanted to interview.

We might have interviewed even more ranch women, but we, too, were ranch women. Ranch women interviewing ranch women was challenging. In between the taping, the typing, the sharing, came the winter feeding, summer branding, cowboy poetry, new grandchildren, households, and careers. Time was slipping by. We had to sort out biases and misconceptions, and open ourselves to the diversity we found. Were the interviews representative of age, of experience? It came time to talk deadlines. We'd have to go with what we had, it would be the best we could do. We regret that we couldn't seem to arrange a convenient time to interview some of the women we had contacted about the project.

Who were these ranch women? Most of them were women we knew. Most of them were like the women who took part in the Cowboy Festival presentation. Some were partners with families, others had their own places, some were once city girls, some had jobs in town and daily ranch chores at home, some had a college education, others were educated in a corral. All of them had a special relationship with the place where they lived and with their livestock. Yet in some way, each story was unique.

Several themes dominate the stories. Some stories, such as those of the Wearmouths, emphasize the legacies of generations, the importance of carrying on the ranching tradition. Place, the land, is paramount in many stories. Others stories concentrate on human relationships, as in the Lawrence and Beierbach stories. Within contemporary families, most stories are positive, but a few are fraught with difficulty. The stories are also about choices – making decisions, maybe to break out of an abusive relationship – about retirement, about juggling a career with the work on a ranch. Always there is the work, the ways in which these women contributed, the ways in which goals were achieved. All of the women are concerned with the future of ranching, especially with the BSE crisis, which is greatly impacting the ranches as this book nears publication. Some fear that ranches will be broken up into acreages or summer residences for urban people, while others worry that huge corporations will take over. The concern for ranching goes hand in hand with the concern for future generations. Most emphasize the ranch as a wonderful place to raise children.

In the stories, especially the generational stories, the reader will discover the many ways in which women's lives have changed. Certainly technology and science have impacted the lives of ranch women. Where some women only went to town once or twice a year in the early twentieth century, others, like Lyn Sauder, now travel distances of fifty miles or more to work off-ranch on a daily basis. Sometimes the women have no choice; they work to help finance the ranch. Other women work for self-fulfillment. All of the women interviewed have more choices than their mothers or grandmothers had, significant choices such as deciding how many children they will have. All of them have more time for recreation and leisure. Most of them are more involved in outdoor work; they can and do calve out heifers, and they are in the corrals on branding day. At the same time, their husbands or partners help out more around the house than their fathers or grandfathers did. Few women were involved in the business decisions of ranching one hundred years ago, but by the 1950s more of them were becoming involved. On many ranches today, women are the main bookkeepers and they are part of ranch decisions. Some changes are less tangible than others. Many women feel they "own" the ranch, it is more to them than just a place where they live with a husband, the way it was for many of their grandmothers. Many of the women feel they have gained respect because they are more involved in more aspects of ranch work and have proven they are capable of tasks most of their grandmothers never thought of doing. Finally, these women are open, candid, and willing to speak out. They exercise a strong sense of self-confidence.

In editing the stories, we decided to condense the commonalities as well as the unique aspects of each story. We tried to retain the grammar, the expressions, all that identify the oral speaker as a ranch woman. We discovered that a "field" in one region is a "pasture" in another. This gave rise to a glossary at the end of the book. We wanted to let each of the women have her own voice. The reader will discover a diverse range of experiences, opinions and beliefs.

Finally we organized the stories in a geographical framework. The framework represents the ranching communities where the interviewers reside: Calgary foothills, Pincher Creek, the Cypress Hills and Eastend, the Wood Mountain uplands, and the Parklands. We have come a long way since one of the early meetings in Medicine Hat when some of us wondered what shape the book would take. We found that a book takes its own shape; it evolves as it progresses.

Along the way, so many people helped us toward our goal. We would like to thank Candace Savage for a workshop on interviewing at Maple Creek. Janice Dickin of the University of Calgary Press provided hours of consultation and insight. She prodded us to keep going when we were lax and encouraged us to stay focused. We also thank Joan Barton and Karen Buttner, who guided the last phases of the book, kept the embers burning. Our families offered encouragement and patience. In particular, we thank Perri Poirier who guided us through the computer challenges. To Mobil Oil, thanks for the initial funding; it enabled us to hide out with the manuscript for weekends at Spring Valley Guest Ranch near Ravenscrag, and at hotels in Medicine Hat, and Lethbridge.

Would we do it again? Who knows? We sure hope someone will. After all, somewhere a ranch woman is putting on her parka, her overshoes, and a scarf and walking out into a stormy night to check the heifers. Maybe she'll have to pull the calf, maybe the heifer will prolapse ... maybe it's another story.

Thelma Poirier

ROCKY MOUNTAIN VIEW

Chestermere, Airdrie, Irricana, Acme, Carstairs, Cochrane, and Bragg Creek are towns in Alberta that have much in common. They all are within view of the majestic Rocky Mountains. Their winters are modified by warm chinook winds which flow over the mountains from the west. All came into being after the arrival of the North-West Mounted Police and the establishment of Fort Calgary in 1875. Chestermere began as a recreation village, but the others were established as centres for the fledgling ranching industry. This industry flourished after the Canadian Pacific Railway and its branch lines were completed in the late 1800s. The land was open range prior to 1905, when Alberta became a province and the Homestead Act was introduced.

Cochrane was named for Senator M.H. Cochrane, who established the Cochrane Ranch there in 1878. It was the first big ranching company in Alberta. The topography is designated as foothills, elevation at the town site is 3,760 feet above sea level, soils vary from grey wooded to black.

Bragg Creek is a similar foothills town farther south and west of Cochrane. This area offers the unspoiled beauty of pine forests and swift streams.

These foothills have a history of sudden summer showers and heavy spring snowfalls, but despite the challenges of weather and of predators such as bears and cougars, many of the ranches here have continued in the same families for four generations.

East and north of Calgary the foothills give way to the gently rolling prairie which encompasses Chestermere, Airdrie, Irricana, Acme and Carstairs, with the Rockies for a western horizon. The names for Chestermere, Air-drie, and Carstairs were taken from locations in Scotland, while Acme is a Greek word meaning "point of perfection." The name "Irricana" is a combination of irrigation and Canada. Elevation is down to 3,000 feet above sea level with annual precipitation of seventeen to twenty inches. Because of the variance in soil types, from sandy loam to black loam, the lower elevation, and the availability of irrigation in some areas, a variety of crops are grown, while grazing is still carried out on land unsuitable for cultivation.

For many years these seven towns were sleepy, rural farm and ranch centres, but during the past thirty years they have evolved into large bedroom communities for the cities of Calgary and Airdrie. Calgary's growth, stimulated by the petroleum industry, also accounts for the many five and ten acre holdings owned by people who work in the city but prefer a rural lifestyle. During the past century, ranch women have watched their region change from open range for livestock, to fenced pastures, irrigated farms, to urban sprawl.

VERNICE WEARMOUTH

My Grandmother Towers would have her babies out in the trees in the back yard, then she'd go back into the house and cook for all the riders that were riding through.

Vernice Wearmouth was born in June 1920 on the Wineglass Ranch, where her Grandfather Towers homesteaded in 1885 near the western edge of Cochrane, Alberta. Vernice and her husband still live there. This historic ranch is now about one-third the size of its original eleven sections and, like many other ranches here, it is continually threatened by urban sprawl from Cochrane. Vernice married Hugh Wearmouth in 1941 and they raised three children, Doug, Renie, and their youngest daughter, Edith, who currently operates the Wineglass Ranch.

I DON'T REMEMBER much about my grandparents really, just what my mother told me. My grandmother Towers would have her babies [five in all] out in the trees in the back yard, then she'd go back into the house and cook for all the riders that were riding through. They had a big room downstairs, and she'd come down from upstairs in the morning and count the bedrolls and make breakfast for them all. And she made all

EDITH WEARMOUTH, LORI-ANNE EKLUND, AND VERNICE WEARMOUTH PREPARING SUPPER FOR THE BRANDING CREW

their overalls and baked all her own bread and everything. She didn't work outside too much. She had all these kids to raise and the house to look after and the men to feed.

When I was growing up on the ranch, we had six or eight men. I used to cook for them and help Mom with the cooking – I'm an only child. And we had a hired girl too. Mom was responsible for milking about eleven cows and she made butter and sold eggs to bring in grocery money. Dad looked after the outside help, the hired men and the haying. It was mostly haying then, very little crop.

I was away at school most of the time. There were very few roads. There was a dirt road through the field from the ranch, and there were no buses. For a while we did go by team or tractor out to the main road, and a neighbour fella drove a car and took some of us to school. I boarded with my grandfather and my aunts for the first four years when I went to Brushy Ridge School, and then I went to Cochrane for four years and boarded in there at a boarding home. Then I came back for two years and went to Springbank because our school at Brushy Ridge burned down. I took the last two years in Cochrane and I boarded in there, so I wasn't home much during my school life. I finished grade eleven and then I went to Agriculture College at Olds. I did two years there. Wherever I was at boarding school, especially the last years in Cochrane, I took piano. It was just something I liked to do, and when we were on the ranch I'd play the piano at night and all the boys that worked for us would gather around the piano and we'd have a singsong. We had to make our own fun and we loved singing around the piano – that led me on further in my music.

I know my parents wanted me to go to agriculture school, but I don't think they had any other plans for me 'cause in those days usually when school ended you came home and got married, you know? I might have gone further if I'd had a choice, but Mom and Dad didn't have the money to send me any further. At one time I thought I'd like to be a nurse, but there was no way they could have afforded it. I know they wanted me to be very honest, kind, a good neighbour, and trustworthy. Dad never set foot in a church and yet he was the most honest, respected man in the country, and Mom was very much that way too. We never started going to church 'til after I was married. I don't think it would have been easier if I had been a male, except with a male you'd have figured your place was on the ranch and there was no question.

When I was home in the summertime, I used to run the rake and the field team, and a girlfriend and I tried driving the sweep one summer, but that was pretty hard because we each had a team. She had the slow one because she was a city girl, and I had the colt on my team. It's funny we didn't break our necks, but we did it for the summer anyway. I was never paid or given allowance, but I guess I was paid with a home and good food. But I always drove a rake out in the field. And when I got older I cooked for the men and did their washing.

I was never, ever allowed at the barn when they were branding or when a cow was calving or when the bulls were turned out. I remember one time a cow was calving over in the shed and Hugh was away and Doug was helping her – I went over and asked if he needed some help, and he said, "You can help me but don't look." I don't know why, it was just a thing that Mom and I weren't allowed. I don't think I've really seen a calf born yet. I have from a distance, but I was brought up like that and you didn't see it.

We had six to eight men who lived in the bunkhouse and came to the house for meals, and I used to get up at five some mornings. This one fellow said he was going to want breakfast before I got up, so I beat him – got up at quarter to five and had his breakfast on the table at five.

Momma always had a girl work for her in the house, and Dad usually had a girl in the hayfield. He said they were far better help than the men – they were a little more patient with the horses. I think to have a girl outside maybe at that time was unusual. I was pretty strict with the girls in the house. I used to hire them and they'd come out to Cochrane on the night bus, and if they didn't shape up by the next morning, I'd put them on the bus and send them back. I got pretty good at hiring and firing girls.

After we were married, I worked out in the field when I was needed. I never did learn to milk cows, which I was glad of. Mom did all that, but we had chickens and turkeys and geese and ducks and milk cows and cattle and horses. We were very self-sufficient.

For the first two years of our marriage, we lived at home with Mom and Dad. A young couple just married living in the next bedroom to their parents wasn't much fun. Later Mom and Dad lived in the house just across the yard.

Hugh never helped in the house that much 'cause I had a hired girl and I didn't need him. He would have if I'd needed him, but I never needed him from the time the kids were born until they were gone to school. I had six pregnancies, and three living. Except for riding with my dad and with Hugh, I didn't really work outside after that 'cause I had the children to look after and the household, and I was still feeding six men and putting up lunches every day for the field. I never had a decent horse 'til later on and I got a real nice horse, so I used to like that.

Our oldest is Doug – he's retired now from the RCMP and living in Sherwood Park at Edmonton. He still does undercover for the RCMP. He gets a disability pension because he got beat up in a bar in Fort McMurray one night. He lost his sight and hearing – his sight came back but his hearing never did. He showed some promise of ranching, but he didn't really think too much of the cows. It was okay, but he'd sooner repair a tractor, and he'd sooner take the bike to go get the milk cows instead of a horse. I think we had hoped that he might, 'cause being the only son we thought that he might take it over. If we hadn't taken it over from Mom and Dad, I don't know what would have happened. I guess it would have been sold.

When the children were small, my day was usually from 6:00 a.m. to about 10:00 p.m. I didn't like housework; I'd just sooner be outside. Just being out in the open was the best. Cleaning out the chicken house, that was the worst part.

I've never worked off the ranch. I don't think it's right. I think mothers should stay home and raise their families. I think that's why there's so many young people in trouble is there's no family life. No family values anymore. We worked together and we didn't buy a thing unless we could afford it. We just did without. I used to wear second-hand clothes. A lady in Cochrane had daughters and they were quite well off and they had nice clothes, and Mom would get the clothes from them and they'd be tailored to fit me.

My children's life on the ranch was quite different from mine in that they had more company. They were at home in their school years and they had their friends over. I never had any friends. I had the odd girl come and stay, but I never really had a close friend, ever, until about six, eight years ago, and then I had a friend that lived across the road. We were just too isolated – I was quite lonely at times. I definitely think a ranch is a good place to raise children. It was always my dream to have the children involved in this ranch. I didn't think it would happen with the girls because they'd get married and move away, but I'd hoped it would happen with Doug. It didn't, but Edith's been the one. You couldn't ask for a better person or a more involved person than she is in ranching and community pasturing, and she looks after soil and grasses and so on. And she's got a daughter that's maybe more interested than Edith ever was. Lori-Anne is twenty, and she's very interested and very involved. She's getting right into it and she's very knowledgeable and enjoys every minute of it. So I think there's been four generations of women, and I wouldn't be surprised if she's the fifth to own this ranch and work on it. I think it's just the love of the outdoors and being able to get out and be with nature and the outdoor freedom.

Hugh is very independent and so am I. He won't even ask Edith for help. It's only by chance if she comes over and sees something. And we thought of moving to Cochrane at one time, before we built here on the hill, but Hugh would have died in there. After we moved up here I became less involved in the ranch. Hugh would go down every day to the ranch and he'd work all day with Edith every day but Sunday, so I was left here. I could have gone, but at that time my house was more important than anything, keeping a tidy house. I've learned that isn't so important anymore. I sometimes used to say, "Well, I can't go riding today, Hugh, 'cause I've got to wash the floor," and Mom used to say, "You don't say that, you go, the floor will wait, but your husband won't wait." I go down for Edith when she's branding. I make twelve or fifteen pies for the branding and then I go down and help her get the lunch and supper on, but that's it. This year I cheated; I got the church ladies to make the pies. I think I've made my last pies for branding.

We plan to stay here in our home on the ranch as long as we can, but the town of Cochrane wants to take half the ranch into the urban fringe. It'll go across this hill to the west of us and across that hill over there and it'll cut the place right in half. And we're fighting that. We went to the Cochrane town council meeting two weeks ago last night, then Tuesday afternoon we were in at the M.D. of Rocky View trying to get it stopped. The only way they can take it is expropriation, unless we decide to put it into nature reserves. I think it won't happen this time but the next time it comes up, it might.

I don't see how they can just take it, but they can. You know, it doesn't make sense. They're taking all the good farm and ranch land, and I don't know how on earth they're going to feed all these people if they keep taking all this good land. We're the buffer zone to the west of Cochrane – if we go, they'll go on right to the Indian reserve. They are sorta nibbling at us. It's like a creeping disease, you know, it just sorta creeps. And I hope I'm six feet under when I see this place go.

It used to be that you never saw couples on a ranch split up. I think it was because they shared a common goal and they worked together and they shared everything together. I think that's the problem now – there's too much else for the women to do. There was nothing I could do except be a ranch wife, but nowadays they have a job in Calgary or a job in a town, so they're not home and they don't share with their husbands. I don't think more recognition for unpaid labour would help. To me, they are restless and they don't see working that hard to make a living on a ranch or a farm. I'm sure that my kids couldn't work as hard as I did in those days. Edith is a very hard worker, but they have more outside social life than we had. The ranch was our life and we worked, and there was no time for fun until after the work was done, but now the fun quite often comes first and the work comes after.

We have a busy social life now. We square danced for years and when our kids were in 4-H we got into extra activities, and when they were in high school I used to drive basketball teams to Beiseker and other places 'cause Doug was quite a basketball player and we were mainly very involved in the United Church. I was secretary of the UCW and the Board, and Hugh was on the Board. We're not quite so involved now, but I taught Sunday School and I sang in the choir and played for the choir and I played for the church. I had fifteen years of music in the church and twelve as organist.

Edith's oldest boy wants to ranch and he's talking about coming home when Edith can't do it any longer. But with Lori-Anne, it all depends who she marries. We're so close to town now, I've suggested Edith sell out this place and buy another place and have money left over. But she said, "I'll never leave here. My roots are too deep. I don't want to leave here."

City people find that very hard to understand. They don't get it.

I don't think the future of ranching is very good. I think there'll be a big shortage of food unless they have these compact places like feedlots and such instead of this open country. There's a friend of ours, her husband's in real estate in Cochrane, and he says in ten years he can see this place gone, as we know it. Well, it's gone now as we knew it, but he says he can see it all developed. I hope he's wrong. I maybe shouldn't say this, but it used to be we were respected and then it went the other way, we were called country bumpkins and they didn't have any respect for you. Now it's equaling out a bit, but we have had quite a few problems with trespassers because of the creek down here. One fellow said, "I have as much right to walk on this land as the person who owns it." So I chase them out if there's a need to, but the RCMP have told me not to because they said they could follow you home. They said this road, Towers Trail, is the worst road in the area for crime.

I've been so happy on this place. It's been tough going sometimes, but I wouldn't change a thing. Except maybe I think I might ride a bit more and I might get more involved in the business end of it, but other than that, I wouldn't change a thing.

I have a lot more community, social life than my mother and grandmother. It's a lot easier with all the modern appliances, but I think Mom had maybe a better life than I did because she wasn't always under pressure to be out running around in the car somewhere. I don't know what the role of ranch women will be in the future – the way things are going, maybe there won't be too many ranch women. We were the majority and now we are becoming the minority. It's a hard life but it's a good life.

Hugh Wearmouth died in February 2005.

EDITH WEARMOUTH

The way I look at it, humans can't go out and eat the grass, but the rancher can change the grass or grain into edible protein.

Edith, the daughter of Vernice and Hugh Wearmouth, is the owner of the historic Wineglass Ranch, where she was born in 1953. The ranch buildings are located on Jumping Pound Creek, a few miles west of Cochrane. Edith started out as a nurse, then married and raised two sons and one daughter. She and her husband took over the ranch from her parents, but she has been in charge alone since her divorce in 1994. Edith is an avid environmentalist and has been recognized by the Alberta government for her efforts.

THE RANCH WAS FOUNDED in 1885 by my great-grandfather, Frances Towers. That was my mom's grandfather. He came over from Birmingham, England at the age of sixteen on a cattle boat, and he got work on the CPR. When he got to Toronto, he met a lady by the name of Elizabeth Glover from Guernsey Island. He married her and he left her there and came West, working on the railroad. When he got to Fort Calgary, he suddenly remembered he had left his wife back there. So he sent for her, brought her out, and he got to about Mitford, just a bit west of Cochrane, and the foreman said to him, "You have acquired some cattle now, and the other end of the tracks is going to meet us soon, coming through the mountains. Your work will soon be done – why don't you homestead some land?" So he did, he homesteaded here in 1885. We have no idea how they settled on the wineglass for a brand, but it came into being on April 28, 1889.

When we were growing up, my brother helped a lot outside, and I remember some summers him actually living in the bunkhouse with the hired men as opposed to living in the house. He drove tractor, mostly the machinery work, and my sister and I did a lot of riding, and just a bit of tractor work. My dad and my grandpa were very strict in that they didn't allow us girls or their wives to be out with the men, so my mom never saw a calf being born until she was well into her sixties because her father and my father wouldn't let us women be out with the hired men. I suppose it was so that they got their work done and so that as girls growing up, he was worried about what was going on, and they didn't want us exposed to the spitting and swearing and so on. In fact, to the point that they would usually saddle our horses and bring them to the house, and then we could get on the horses. We didn't even go to the barn and saddle our own horses. We could, but if the hired men were out there, they didn't want us to be.

The first three years of school I went to Brushy Ridge School and all the kids there were from ranches. But when I got to Cochrane it was different. At that time it wasn't cool to be from the country. Now everybody wants to be a cowboy, and dresses western and Wranglers are the "in" thing, but back then it wasn't.

I was very, very surprised that this is the occupation that I have ended up with because I actually took nursing right out of grade twelve and never thought I would end up coming back to the ranch. I think my parents had hoped that somebody would take over, but I don't think they expected it. Now there are so many things kids can go into, but then it seemed that the choices for girls were teacher, nurse, or secretary. So, I was a nurse. When I went away to nursing, I had no idea where that was going to take me. Then we got married and we came back here to work for my mom and dad, and here I am. That decision would have absolutely been easier if I had been a boy. I think everybody was pretty happy that I was going to do this, but it certainly is a male world, or it was a lot more so then.

We were the boss's daughters, and so Dad would give us the dirty jobs, but he would give them to my brother too. Like if the barn needed cleaning out, we all got to clean it out. Or like driving the tractor, or with Mom, she had a lot of hired men, we would have to help her make the lunches every day for the hayfield. She'd have sometimes eight or ten lunches to make, and we'd have to help pick berries and stuff, but it wasn't just because we were women – it was just because it needed to be done.

Mom tells the story that some of the hired men they had would help dry dishes after supper. This one fellow refused to dry dishes, so they tied him to the table leg until he started to help in the kitchen. Some of them wouldn't clear the table, I remember, and that really got my dander up. This was when I was cooking for the hired men, it was only two or three at that time, but I think they soon changed their ways.

After we were married and we came back to the ranch, I assumed the role that my mother had, so I cooked and cleaned for the hired men. They either had a bunkhouse or a mobile home. Once a week for the single men I would clean their accommodation and do their laundry. And I cooked their three meals a day for them. I had the chickens on my own, and we made our own butter and our own bread and things like that. I also tried to ride and drive tractor as much as I could, but that sort of faded off as the kids came along.

As the children grew, they all rode from a fairly early age, especially the boys – the two oldest, they were with their dad and helping fence or fix machinery. Later on, if there was a cow calving or something, we'd say, go and bring her in. They had the opportunity to go with him all the time.

I didn't treat my sons and daughter the same, absolutely not. Unfortunately, my sons say, "Mom, you really spoiled Lori-Anne," and I'm sure this is true. I really don't know why. I guess we thought they were older, tougher, and she was more fragile or something. Looking back from the position I'm at in my life now, I'm thinking that was really strange. But at that point in time, that was how it was for the women, particularly on this place. They weren't sort of involved and they weren't treated the same – it wasn't good or bad, it just was.

Everything was decided by the men, very much so, absolutely, and who knows why that was the way it was, but it was. I think it was probably the way it was with Mom and Dad, as far as the ranch actually went. I didn't know anything about revolving loans or much about banking or anything. The men made all those decisions.

I remember every morning I'd get the kids up and get them dressed, and we'd have a big breakfast – there was either one or two hired men when the kids were little. After that we had married men, so they didn't come here to eat. Then I'd probably do the chickens, take the kids with me. Then I'd strain the milk – the men milked the cows. I baked bread once a week. Mondays I'd do laundry, Tuesdays I made butter, Wednesdays I made bread, can't remember what I did Thursday, maybe went to town, Friday was always cleaning day, Sunday was always church and it still is. So then I'd take lunch out to the field or send it or they'd come in here. They'd usually come in for morning tea or afternoon tea, so I'd always have cakes or cookies, and then we'd have a big meal for supper, either in here or out in the field. Then when the kids were in school, I'd go to the school for their extracurricular things, and riding when I could. I just always felt like I was looking after some body else.

I never thought about going to work off the ranch full-time – everything I want is here. And when I do go to work for a day or so, it's nice to get dressed up and see other people, that's what I really like. I find myself in jeans a lot and grubby a lot, and so I really enjoy a day at work, but everything I could possibly want is right here. When some women with young children go out to work, maybe they need to financially or for their sanity. I think it is good if they can stay at home, but I certainly don't judge anybody who chooses not to. I was very, very fortunate that both my parents were here all the time when I was growing up. My kids saw both their dad and I a lot when they were growing up.

We always had a huge garden. We did a lot of canning and freezing. When I had the kids, I guess we bought virtually no groceries. I guess we bought flour, but we bought no meat, no vegetables, and no milk, canned all my own fruit, froze everything.

I love my kids to be involved and they come out. But I only want them to be involved in the business of ranching if they want to, and I've tried to set it up and I've told them this – if none of them decide to come back and ranch, then that's just fine. I think kids on a ranch have a great sense of where they belong, and a great sense of family and a great sense of community.

I told my kids they're not getting me off here for a long, long, long time. But I can see at some point in time if one or more of them is interested, then we'll have to make that transition and I will step back. Which, I'm sure, is very hard. I don't know how my mom and dad did it, as they made the transition over to me. My dad is still very, very involved, but to step back and not to still run things.... If my kids don't want to take over, then I will just continue as I am. I hope to be like my dad, at the age of eighty-three still be riding and fencing and possibly driving tractor, if I can figure out the newfangled tractors. I have some arthritis now and a bad back, so I definitely can't lift things the way I used to.

When I found myself running the ranch on my own after my divorce in 1994, I thought I'd have to learn a lot about fertilizer, and when to take a crop off – I kind of know when, but my interests lie more with the animals, so I got in this joint venture with my neighbour and I sold my line of equipment. And my dad helps – he's not hired but he works here.

It's very different now, because I guess I have more confidence now to be a social person, and that's just come in the last few years, as well from me being on my own. Before I was somebody else's wife. And then I became suddenly not somebody's wife anymore, and it was just a very extreme learning curve. We, as females, had never had much knowledge of the inner workings of the ranch, and then all of a sudden I was in charge. My friends and I go to the theatre a lot, we go to the philharmonic – I love that sort of thing, just a big variety of things, our Cowgirl Cattle Company. My church plays a large role in my life.

I'm involved with Cows and Fish, and I'm currently on the Ag-Service Board for the M.D. of Rocky View, Alberta Cattle Commission, and Western Stock Growers. I've had a lot to do with Trouts Unlimited and their work with ranches. It's very, very important for women to get involved with these organizations. When I first came into this ranching business on my own, I'd go to a meeting and I'd be the only woman, and quite often I still am – in fact on the Service Board I'm the only woman. And when we meet with the ranchers, I'm still the only woman, unless some of the wives come along. And what I observed in some of our meetings, when there's a difference of opinion, that watching the men is kind of like a cock fight or a rooster fight, you know – they get fluffed up bigger and bigger and bigger as it gets a little more tense. And I'm thinking that women have a lot to add, not just in calming things down but keeping a reasonable head. I can see the men getting aggressive, whereas the women might say, "Okay, let's figure this out." They bring a sense of calmness and a sense of, maybe, a little more fairness. They're not as threatened. So in these organizations I think it would be very beneficial to have more women to bring a balance. Before the divorce, it would always have been my husband who was involved.

We're busier than Mom and Grandma, but they worked harder. By busier I mean we're busy with things that maybe aren't quite as important, too many demands on our time. There's more stress now. I think we bring stress on ourselves, so it's a bit of our own doing, and I think a lot of it has to do with living in an affluent society in this part of the world. Back then they were too busy to worry about little details. I also think that now we work with so much red tape and so many laws in every aspect of our lives. And I know in my ranching life right now I spend well over half my time just doing managing – in fact, last week I said to Heather, the girl who works for me, all I want to do is go out and check the cows, and I could not get out of here with so many other demands on this ranch being so close to town. That's something my grandparents never had to deal with – they could get on with the business of ranching. And this is still the business of ranching, but it's become so complicated. More business than actual physical ranching.

One of the skills women need to have is the willingness to learn more, and that's one big thing that the male gender quite often don't have, that willingness to learn. There were a lot of things I didn't know. I knew how to chase cows and I knew how to check grass and I knew how to judge cattle and I knew calving problems. But I didn't know anything about banking or interest rates or the business end. And I think for me it was about getting a good team of people. My accountant, my financial adviser, a lawyer, a neighbour I could phone up and say, "I don't know anything about barley, we have to plant some barley, how do I go about this?"

And I think a ranch wife needs to know more now to be a part of a team. In the future I think ranch women are going to have to be far more business-oriented. I've told all my children that if they think they ever want to come back here, the very least they have to have is two years at Olds College because there is so much figuring and so many numbers and so much business end of it that you have to have at least that. Cochrane is getting bigger – we're going through some annexation stuff right now – so they would need to take some public speaking courses or some Dale Carnegie courses so they can stand up and make a logical presentation in council, which I had to do last week. So I think if my daughter was to take over, these are some of the skills she would have to be very versed on, even more than how to ride or how to fence. The girl who works for me can ride and fence way better than I can simply because I've gotten away from that because I've had to do more of the business end.

The future of ranching in Alberta is scary – very, very scary. I think until we can educate the general population, the voting population, as to where their food is coming from, they'll just keep gobbling up the agricultural land, and we're going to need some kind of legislation from our government officials. I think the small ranchers are just going to be squeezed out, and the big land barons will own the land and just be absentee ranchers. Everybody wants to move out to the country and then close the door after they get here. Somewhere somebody's got to have the guts to pass a law to change how much agricultural land is being gobbled up for building. The way I look at it, humans can't go out and eat the grass, but the rancher can change the grass or grain into edible protein. It won't change in my lifetime, but hopefully we'll start something and then someday we'll save some of these ranches and farms.

I'm seriously looking at a Nature Conservancy for this place. I've been quite involved with the Southern Alberta Land Trust helping to get it off the ground. I have to look into it a little bit more then – it's still fairly new in Alberta, and it's still going through some growing pains, so I want to make sure I look into every aspect of it. That could be the way, but I'd hate to see the whole country put into land trust, in case something does go wrong, in case there is some glitch in it and some lawyer finds a hole in it and then we have a big mess. But I'm sure looking at it for this place. Next week I'm going to a range management seminar, and I think as ranchers we have to really be careful of the grass. The grass is our factory, that's our business. Hopefully we learn how to manage it a little bit better all the time.

To young ranch women starting out today, I'd say make sure to do things on your own. Make sure you have your own life. Go into something so that you can support yourself. If you get married that's fine, but make sure that you have something so that if you end up on your own you don't have to take a minimum-paying job. I just remember it being very difficult when I first started on my own. It's definitely a man's world, and I remember even one time I phoned somewhere for parts and I didn't know much about parts. I tried to explain it, and finally a fellow came on the line and asked, "Is your husband there? Or anybody else we can talk to?" and I said, "No." I remember being in tears for days over that. There were quite a few incidents like that while I was figuring things out, but I became more confident. But it was a huge time of learning and growth for me. One that is still a work-in-progress.

For a girl it doesn't matter if it's a ranch or what, but just make sure you go out and make your own place in the world and be your own person first. Most of us have always been somebody's wife, somebody's daughter, and somebody's mother. One of the best things I remember is getting a whole stack of Christmas cards just for me. I was my own person.

I feel so privileged to be able to ranch and to have the opportunity to do what I love. This is the only place where the world makes sense. When I'm surrounded by nature, the world feels abundant, caring, and generous. My mind feels clearer, things sort themselves out. Peace, solitude, a deep connection, deep roots, a sense of belonging, of being grounded come from this place. The natural environment and people's connection with it are disappearing more and more every day, and I believe as nature disappears, the mental, physical, and spiritual health of society slowly disappears as well. I believe that the farmers and ranchers who manage the larger tracts of land in the future will hold the key to a healthy human race.

Edith Wearmouth recently received an environmental award from the province of Alberta for her work in the improvement of Jumping Pound Creek and in maintaining her grassland.

HEATHER MCCUAIG

I'd love to have my own place.

Heather (Harden) McCuaig grew up on a stock farm south of Fir Mountain, Saskatchewan in the 1960s. She was the fourth child in a family of six children. After hiring on at the Wineglass Ranch at Cochrane, Alberta, she met and married Greg McCuaig, a locksmith. She has two children: Colton, age five, and Cody, age two. Heather and her family presently live at Caroline, Alberta. She was working at the Wineglass Ranch when the interview took place.

I THINK IT'S A GOOD TIME to be working as a ranch hand. I don't make millions at it, but it's a way of living. I love the freedom of being out there and the simple things in life – you know, watching the cows feed, training horses. And training dogs.

I always thought I would get a good-paying job, training and working for the government somewhere as a nurse or teacher, but you know, going to school and taking my pre-nursing classes in Medicine Hat, I just felt like that was not

THE MCCUAIG FAMILY, READY FOR A ROUND-UP –
GREG AND CODY, HEATHER AND COLTON

for me. There was something better for me. I realized when I was studying the books and being in classes that life was too short to learn something that I was not enjoying.

When I was growing up, I followed my dad wherever he went, always enjoyed what he did and what he taught me. I was driving tractors and trucks and working cattle and so forth at home. I just loved the outdoors. And then I went on to Agribition and was clipping cows and getting them ready for show. And yeah, I enjoy riding – I get on my horse and I get totally lost. You know, I'm not near a phone and it's a getaway for me.

I always dreamed about coming out west, saw pictures with the mountains in the background. Dad was friends with Lindsey Eklund, and he said his hired man was leaving, and I just said I'd like to come and work, and here I am. I've been here for nine years, and I enjoy it.

Here at the Wineglass there's Edith. She doesn't do the manual labour like I do – she takes care of the book part of it. Then there's Hugh, that's her father. He's eighty-five and he's a big part of it. If there's riding to be done, he's out there riding. We don't leave him behind because he'd feel bad about that. Since Lindsey and Edith split up, Edith can't afford to go out and buy a bunch of machinery to run her property, so she worked out a deal with the Quarter Circle X Ranch run by the Buckleys, and they farm Edith's land. It saves buying the machinery, which nowadays is pretty costly. It works out well.

My husband has his own business in town, but he helps out when I need him, like at calving season he does come out and give me a hand if I have trouble. A lot of times I am by myself and I need that other person there, like if I'm pulling a calf, maybe to pull the cow sideways, you know.

Everyone always asks Edith, "What does Heather do with her kids?" Well, they come with me, they've always had to come with me. They entertain themselves quite well. If we're out fixing fence or if I'm rolling wire, they run in front of me and we make a game out of it, always try to make a game out of it. Sometimes they get bored, but they've always had to come with me. They just know and they just hang out. Edith tells everybody they just hang out, and people say, "What do you mean they just hang out?" They just do. It's pretty safe, though we have a creek running here, but it's always been drilled in their minds – don't go next to the creek unless there's somebody there. There's always somebody close by, and yeah, it's like anything, you take risks, and I just hope if something ever does happen to my kids, I hope that I'm close by to be there for them. Yeah, they just come with me and they just know. They come on the tractor to feed – it's pretty crowded nowadays in the cab, but yeah, we do counting, we do our ABCs and stuff in there – it's quite fun.

Oh yeah, there are times when it's very difficult. Until you take your kids with you day in and day out, a person doesn't know – like it's different when you work for yourself. In my position, I feel like I'm committed to someone else, like I got to do that job. If you're on your own and running your own business, then you can just say, oh well, I can do this another day. Yes, I guess I get frustrated, but it's a learning process because I'm not the most patient person and it's taught me a lot of patience.

Just this last year Colton was starting to get into this riding stuff and he really enjoys riding. But anytime I'm doing big rides and I have to be there with my dogs, they don't come with me. I don't want to sour them on that, I don't want to push them past a limit.

I got the cell phone when I just had Colton with me and I was pregnant with Cody and I was feeding on the backside – I had to carry Colton down the hill twice. I got high-centered with the tractor, and the other time the tractor just quit on me.

When I'm stuck I don't want to leave the kids back there and run home, they're just too young to do that sort of thing, so I say, let's get warmed up and start walking, guys. And we make a game out of that, too.

I can move a lot of cows just myself. I'll be three-quarters of a mile behind and my dogs will be at the back, or I'll send the dogs to kind of turn them and stuff. You get to that point where you don't know what you'd do without the dogs, really.

I've always loved dogs. And when I came out here, my brother Jody gave me a pup. The pup was born the first of November and I started the first of February, so she was only three months old. We got quite an attachment and now she's nine and I got her working. I said to my brothers, "How do I train her?" but she trained me. Border collies or any working dogs are so intelligent. She's got a little hip dysplasia, so now I have another pup I got from my brother Ward, and this one's come along really good. They're starting to work together. I've got the young pup going on commands. Yeah, they're starting to click.

In the winter I make sure the water's fine, feed, sort thin cows and put them with a bunch that are getting more TLC, that kind of thing. And then I'm doing mechanical stuff, like maintaining the equipment and getting it ready for spring, and then all of a sudden I'm doing spring stuff, trying to put seed in the ground, and then right after that I'm getting ready for haying.

Since I've had the kids, Edith gets someone else to do the farming. It doesn't hurt my feelings at all because I get to do the cow stuff, and I'm happy doing that and I can do that with the kids. You know, I can put up an electric fence and turn the cows in the fields where they need to be in the summer. Sometimes John will ask if I could do some discing for him, and I say sure, but six hours at a time on the tractor is about as much as I can handle with the kids. It works out pretty good.

I've had quite a few incidents happen, and when I get in a bind I think, how am I going to get out of this? It's kind of neat actually, because I'm kind of hard-headed and I don't like asking people for help, so I just try to work it out myself. John has been awesome and he says, "You know, Heather, we rely on you too much. We want you to be our eyes and our ears and if you need help let us know." And I have called on him for a few things. When I try to go it alone, that's when I get myself stressed out, so we all have a meeting and it puts me back in perspective again.

Both Edith and John Buckley let me be a part of it. They let me have a lot of say and well, it's the next thing to being manager, because John tells me, "You know the cows better than I do, you tell me what has to be done." That's a lot of responsibility to have.

When the Buckleys and Edith got into this venture, we weren't utilizing the grass where we should have been because of the water system – everything watered down at the creek – so we set up a well with a big water trough, a big tank on top the hill. I was hauling water to them. Anyway, we had an old truck called Pink – it was an old Dodge, '63, three-ton. So we put a thousand-gallon water tank in it and I'd pump out of the creek and fill the tank. One morning I was in a little bit of a rush. Anyway, I was climbing up the hill in the old '63. You have to wait until Pink gets warmed up pretty good. Otherwise she chokes and she coughs on you, and to make a long story short, I was going up this steep hill and she coughed and I clutched it and she coughed again and all the weight got sloshing and the front end came right off the ground. And I dumped the thousand-gallon tank and the wheels were pretty knock-kneed. I thought, oh my God, I didn't believe this would happen. Anyway the truck and tank of water was still kind of in one spot. John came out and helped fix it, and he tried to make me feel good by saying, "Heather, you know, you're giving 110 percent," but I was thinking, I shouldn't be having stuff like that happen.

The future? The big problem is urban sprawl. Being so close to Calgary, there's always a fear of being annexed and having the land taken away. I'd love to have my own place, build something of our own, but sometimes money doesn't let a person do that kind of thing. Maybe I'll have to go back to Saskatchewan, move home. I can see that happening. I said I'd never go back, but to be able to afford something out here, you have to be a millionaire.

IRENE EDGE DRESSED IN HER SUNDAY BEST

IRENE EDGE

*Hopefully my great-grandchildren will be able to ranch here,
but it's getting harder for them to stay.*

Irene's story is similar to that of many ranch women in that she was raised in the city and moved to the Cochrane area to teach at a small rural school. There she met and married a local rancher. Irene and David Edge were married in 1940 and moved into the house where she lives to this day, on a hill south of Cochrane with a panoramic view of the Rocky Mountains. She and David raised two children: Brian, a veterinarian and rancher, and Beryl, a nurse, who is married to a local rancher. Irene is the mother of Beryl Sibbald and the grandmother of Cheryl Nixdorff, whose stories follow.

WHEN I GOT MARRIED in 1940, I was twenty-one, and learning to cook on a coal and wood stove was the biggest adjustment, although when I was teaching school I boarded on a ranch near here for a couple years and so I had learned to live in a rural setting. They had no running water or electricity or gas there, so it wasn't too big an adjustment when I first got married. I was boarding with this family and the lady had a baby each year for the two years I was there, and they were born at home. The first one, she had a midwife and a doctor in attendance, and I was home for that one. I was sitting in the kitchen knitting for a while, then I was working on a crossword puzzle, and the doctor would come out and help me with a word or two and then he would go back in. It was a very small house, so I really felt that I was right there. The second one, I was at school and wasn't expecting the baby to be born, and when I got home the baby was there.

When we were first married, David quite astounded me by saying, "Now I want you to know – we'll get this straight right now – I am the boss outside and you can be the boss inside. If I am at the far end of the field, I don't want to have to worry about being in to dinner right at twelve o'clock." And I said, "OK, if I am in the middle of a job and I mightn't be done until five or ten minutes past twelve, don't you worry either if dinner is a touch late." So it was fine and we never had a problem that way. He had the greater knowledge as far as the ranching business went, but we certainly discussed things and agreed on major decisions. We had common goals.

I didn't really do too much outside except riding – I enjoyed riding with my husband when we were moving cattle or things like that. We would sometimes ride with the children, too, but quite often David would ride with them and I'd be left behind.

We were married a year and a half when Brian was born, and that was a lot of work without any running water or electricity, and none of the things we take for granted now. We hauled our water in from his mother's or brother's place on a stone boat. We didn't have a well until after Brian was born. David helped to some extent. He was always very good at helping with the dishes after supper, and he would help me do anything if it was necessary. He always kept the stove going in winter so we could heat the baby's bottle. He was very good with the children, and he involved them in the ranch activities when they were old enough.

When Brian was old enough, he would drive the horses on the hay rake – we did a lot of the work with horses in the early years. And they could drive a car in the field when they were quite young. I always remember Beryl was quite small when they were threshing, and they needed someone to bring a truck in from the field. David had shown her everything she needed to know to drive, but when she got to the garage she had forgotten what to do with the brakes. It didn't really do too much damage, just to the end of the garage.

A typical day for me in haying season – you'd be up about six o'clock, and have breakfast ready for whoever was on the haying crew at seven. David thought they should be out in the fields by eight, of course depending on the dew. Then they would have their lunch in the field because they were often several miles from home. So you would make up maybe five loaves of sandwiches, and you would make a cake in the morning, and have a box of apples and a big box of tea for them. And then in the afternoon, you would prepare their supper.

When Beryl was about thirteen, in grade eight, I had a call one morning about 8:00 a.m. from the principal of the Cochrane School to see if I would go subbing that day. I said no, I had never considered it, and he said to think about it and call him back. So I discussed it with my family, and decided I would try it. On my way in to Cochrane I realized it was April 1st and I thought maybe it was an April Fool's joke. When I got there I said to the principal, "If this is an April Fool's joke, I'll be happy to go home." And he said "No, this is the real thing." So that was the start of my mini-career, which I thoroughly enjoyed. I did subbing in Cochrane School and Springbank School and the odd little rural school around, like Beaupre. And then I was also the first secretarial help in the Springbank School – in fact I was the first secretarial help in a Rocky View school. The principal had a very heavy schedule – he was teaching seven periods out of eight and needed a little help with his administrative things. As I had had a secretarial course, I was asked to work two half-days a week, and it was quite pleasant. Then it became three half-days a week, and then three full days a week. I decided when I had my first grandchild I would retire so I could have some time to spend with her.

So I enjoyed those years and I didn't worry about making a fortune. I often said I would have paid them to let me go because I enjoyed the work so much. David didn't mind – he said whatever made me happy – and it was quite a lot of fun. As I said, I didn't have to make my living, but it was nice for all the little extra gifts I wanted to buy.

We always thought we'd have three children. A boy like David, then a daughter, and then a red-haired freckle-faced boy, just like my father. But that one never arrived. They were born in Calgary, in the Holy Cross Hospital. I just went in and stayed with my mother for a couple weeks before they were born. David said to me before Brian was born, "Now don't make me lose a day's binding." They were binding a crop of oats that was very heavy, on some breaking. It must have been very dry that year because it was ripe on the 21st of August. So I phoned him about 5 p.m. and he was in the house getting a bath, so he came to town and he was there when Brian was born – in the hospital but not in the delivery room. They didn't even consider that in those days.

I think a ranch is the best place to raise a family. I always felt very happy that the children were at home – you didn't feel you had any problems with the children because they were with you all the time. We knew where they were. In the country you have a real sense of community, and if anyone has a bereavement or something like that, you feel it personally for them and they do for you and you find a lot of support in your friends and neighbours – I feel it's real nice to be part of a community.

I had to give up doing a lot of the riding, as I got older. I still always kept the books and David always liked me to be there for any business discussions – not that I partook in them, but he always said that two heads were better than one and I could maybe remember what he didn't, and we would discuss things. Like a silent partner. I still have some cattle, cows and calves, which my son looks after for me, but as far as doing any actual work, I don't have to now. And for the brandings and that, I just enjoy going and visiting and taking part instead of having to do it myself. I'm very happy to have lived here on this ranch for sixty years in the same house we moved into when we were married in January of 1940.

We always enjoyed having people in for the evening, and we'd play cards, usually poker. I always remember people coming on a real cold winter night, and they were bundled up in the sleigh with straw bales and we played cards, we enjoyed that. And dancing, we enjoyed all the local dances. They seemed to have more of them then, and we quite enjoyed them. People didn't go to Calgary to the same extent as they do now. If we went to visit neighbours, the children always came with us. When we curled, we curled in the afternoon when the children were in school, and then we hurried home so we would be home when they came home from school.

Now I enjoy bridge two or three times a week, and I enjoyed travelling, although I think my travelling days are over. I went on a lot of trips and cruises, and I really enjoyed that. I do enjoy the sociability with the ladies I play bridge with. We were members of the Western Stock Growers for a long time, and we enjoyed going to the conventions very much. There weren't many women involved in the meetings then. It was strictly a male-dominated world, but we attended some of the lectures and I always enjoyed it very much when discussions got a little on the heated side. I definitely think that women are taking a more active role in many of these organizations, and I think it's a good thing. We have to realize women often have very good ideas in the management and running of things. I think most women now do have more say in the active management of the places than we did. It is a different era.

In some ways their lives are easier than mine was, but not in others. You know they have all the automatic washers and dryers and vacuum cleaners and things that have come with electricity in the rural areas, which we'd never want to do without now. I think that their lives in the area of housekeeping are a lot easier. Physically easier, but otherwise I think they have more stresses than we did. I think we had a much simpler life. Yes, we did. Their lives are so much more stressful now, but mostly a lot of that is as you make it. You've got to make your choices, and I think that some of them a lot of times choose a lifestyle that to me is too busy. I don't think it's necessary for a lot of them to do that. I don't think that they are expected to do more; I think that a lot of them want to do that. Women are more career-oriented today, and I think that they often want to do these things, but I don't think they are always necessary. I like the mothers to stay home with their children. I think it's so important, those first years, and if you miss it you're never going to get it back.

I really think I've been singularly blessed because I've felt very happy in my marriage and my life. There's very little I would want to change. I'd like to live a lot of it over again, but I figure it's been a good life. All my grandchildren, their parents are involved in agriculture, and they all are living a rural life and I just feel very fortunate that this is so. I don't know what life will be like for my great-grandchildren. This computer age is so different – they'll have access to a lot of knowledge and things, and I'm sure it's going to be quite an asset for them in some ways. Sometimes I think it makes things more complicated for them too.

I plan to stay right here as long as I am able. It's twenty minutes to Calgary and twenty minutes to Cochrane, so I have the best of both worlds. My family, most of them are all right here within easy driving range and I feel very close – I'm in contact with family members every day. I'm able to drive and I'm very happy, and as long as I can I'm going to stay right here. That was one of my ambitions – to see the millennium and then be in my house for sixty years, which was last January.

The city seems to be getting closer all the time, although right here we are fairly fortunate – there hasn't been as much development as in some areas. Hopefully my great-great-grandchildren will be able to ranch here, but I think it's getting harder for them to stay. I think ranching will always be a part of the economy in Alberta; I think it will go on. It may not be the same, but ranchers will expand into other areas.

BERYL SIBBALD MOVING COWS

BERYL SIBBALD

It is satisfying to see that the ranch is going on with our son.

Beryl Sibbald is the daughter of David and Irene Edge, and the mother of Cheryl Nixdorff. She was raised on the Edge ranch south of Cochrane and moved to her present home, twenty miles southwest of Cochrane, when she and John Sibbald were married in 1966. John's grandfather, Frank Sibbald, started the Sibbald ranch in the 1880s. Frank had come west with his father, Andrew Sibbald, in 1875. Andrew was one of the first teachers of the Indians at a place known at that time as Morleyville.

Beryl's home is on a hillside overlooking the valley of Little Jumping Pound Creek, where John's grandfather built his log home. From the kitchen window Beryl and John have an impressive view of Moose Mountain and the Kananaskis area.

The Sibbalds raise purebred Red Angus as well as commercial cattle. Their son and daughter-in-law, Jay and Kari, who have three young daughters, ranch with them.

I WENT TO A ONE-ROOM SCHOOL, Brushy Ridge, until grade eight. My aunt, Eddie Edge, was our teacher for most of those years, and I was the only one in my grade. The school was a mile and a half north of us. We rode to school and there was a barn where we could tie the horses up. In the winter we had a cutter and an old horse, Stoney, and we'd go in the cutter. Or sometimes if it was really cold, we'd get driven to school. I think this was a more secure environment than a private school or a public school in the city. I felt comfortable with all my friends, and even the kids in the bigger grades were my friends and they'd help me. I think a small school had advantages. We learned to work on our own; we didn't always have to have a teacher telling us to work. There were seven grades in one room and she'd give us our assignment and we'd go ahead on our own. When I got to grade eight in Cochrane, I wasn't behind the rest of the kids. I graduated from Cochrane High School.

Mom and Dad always expected my brother Brian, and me to have an education separate from ranching, so we could make our living if we didn't have ranching income. I didn't want to be a teacher or a secretary, and nursing appealed to me, so I took nursing at the General Hospital in Calgary.

When I was old enough to help at home, I helped Mom in the summertime when we had hayers. I would do the baking and then I'd be free to go riding with Dad. Dad had land rented on the Morley Indian Reserve, so we had to go up there quite often and check the cows and move them around. Dad and I mainly did all the riding, which is a nice life.

When John and I were first married, I helped chase cows a lot, too. I've never had to go out and feed or work in the fields haying or harvesting. I've been lucky that way. Mine was just a nice job to go out and help him chase cows. And I worked as an R.N. at the General Hospital and the Foothills Hospital for about three years after we were married. When the kids were about eight, I started working part-time in a doctor's office in Cochrane. Later, I went back and took a refresher course at the Foothills Hospital, and I ended up working for twenty years part-time in the doctor's office in Cochrane. I think it's very important for a mother to be home with preschool kids if she can be so lucky to not have to work. I know some people have to work – it's hard.

John has never helped much inside and I mainly looked after the kids when they were little, but John was always very supportive. I couldn't go out to help as much outside when the kids were little, but I did sometimes go down and help my dad, and Mom would look after Cheryl. Or else we had Ede, John's mom, who could babysit. When the children were about four or five, they started riding with us.

As far as the decisions went, John and his brother made them mainly, but John's dad's brother and sister, Wilfred and Aileen Sibbald, still lived in the old house in the valley. They had never married and they had a say in the decisions too.

Their house was the original log house, which had been added onto over the years. It was right along the creek and it had clapboard on the outside and was painted white. It was always a beautiful house. They had all kinds of beautiful Indian costumes, long headdresses that Frank Sibbald had traded for with the Indians. They had two wolf heads. The one wolf supposedly killed over a hundred head of cattle. The Indians shot him but Frank got the head. They also had bear, sheep, and deer heads. It was a beautiful living room with a player piano, beautiful antique furniture, old pictures, and the family Bible with everything written in it. About twenty-two years ago the house burned down. At that time Aileen was the only one left living there, and she got out with only her purse and her coat and her pyjamas. Everything else was lost. There was a forty-year-old deep freeze in the porch and the wiring or something started it, but we couldn't do anything to keep it, it was just ... gone.

Aileen lived here with us while she built a small cedar log house in the same location as the original house. She lived there until a few years ago, and she lives in the seniors' lodge in Cochrane now. Our son Jay, and his wife, Kari, and their children live in her house now.

It is satisfying to see that the ranch is going on with our son because I think we always expected that Jay would one day ranch with his dad. I think it's kind of unfair for girls, but unless they are very, very aggressive about wanting to stay at home and wanting to be a rancher, I think they are more encouraged to get an education and do something else. The boy can just say "I want to do that," and he can, whereas the girl has to fight really hard if the place isn't big enough to have two come home.

Women's role in the livestock organizations has definitely changed. Women are really very active in these organizations, such as Stock Growers and the Red Angus Association. Good for them. If they have the time and they're interested, I think that's great. There are more women in every organization, especially being presidents and chairmen. Before, all the Red Angus directors were men, and they had one woman, who was secretary. Now women are more involved in every aspect of the organization.

I like this lifestyle and I think a ranch is a good place to raise children because you can keep the kids busy with different activities. Plus, they learn about life, being around animals, the calves being born, and when animals die. I'd like to see my grandchildren grow up healthy and be very involved in ranch activities. I'd also like to see them get an education and be able to do whatever they like. I'd still like them to be involved with ranching and agriculture, if that's what they want, if they can afford it. Lots of ranches or farms are now at capacity as far as making a decent living for one family. At today's land prices, it is extremely hard for parents to buy more land in order to expand and enable their children to continue ranching. The grandchildren may not be able to afford ranching because one place is too small for all to make a living. Also, parents should not be expected to just hand everything over – they also have to make a living.

The plan is to keep going here, if a person can, financially. It's hard to make a living because of such high expenses, and we can never expand here because the price of land is too high. But we'd never sell this place, we love it too much. I've never felt isolated here even though we can't see any neighbours. I think you'd be more isolated in the city. It was hard for Cheryl to leave here and live where she could not see the mountains every day. John and I found it hard too – he cried when she got married.

CHERYL NIXDORFF WITH DAUGHTER ELISE

CHERYL NIXDORFF

I feel very fortunate to be able to stay at home with my kids while they are little and growing up. I think a ranch is the best place to raise kids.

Cheryl Nixdorff, age thirty-one, is the daughter of Beryl Sibbald and the grand-daughter of Irene Edge. She grew up on the Sibbald Ranch in the Jumping Pound area, southwest of Cochrane, Alberta. She is married to Paul Nixdorff and they are part of a family farming and commercial ranching/purebred Hereford operation twelve miles east of Airdrie, Alberta. Cheryl has two children: Brady, nine, and Elise, six.

I THINK THAT LIVING on a ranch was part of me, it was what I grew up with, and where I am now. I never made a conscious decision to go out and marry a farmer or rancher, but I think it was what I was interested in and the type of lifestyle I wanted. And I'm so glad – I wouldn't want to be anywhere else. It's so beautiful out there where I grew up. I could look out the kitchen window and see Moose Mountain every morning. It was hard to leave because it was my home, but I was excited to start my new life with my husband. We are very fortunate to be where we are. We have a nice yard with lots of trees. It's flat, definitely flatter than Jumping Pound, but it's beautiful here too. I love going out there to visit my parents, but I definitely feel at home here.

When I was growing up, I remember going with Dad and my brother, Jay, to help with the cattle. Dad did a lot of the stuff that had to be done on a daily basis, feeding the calves in the winter, fixing fences, checking cows, haying, etc. Mom did a lot of work in the house and around the home, but she did an amazing amount of work outside as well. She was always helping my dad, chasing the cows, getting them in, doctoring the calves when they needed it, running the cows through the chutes, getting ready for sale time and branding.

I was lucky – I got to do both outside and inside work. I loved doing the riding. That was always one of my absolute favourite things to do. And I remember the branding days were just something I looked forward to for months because it was just such a great day. We'd get together and people from all over would come and we'd see kids and go riding and it was just a really good family day. Mom taught me a lot, how to bake cakes and make meals. And Dad taught me how to look at a cow and try and see when they are going to calve and that kind of thing. I don't think my brother was taught to bake or clean, not that I can remember. I think one time when Mom was in the hospital for something Dad had to do his own cooking. Mom told me that when she was in the hospital having me, she was in for a week because they stayed in that long then. When she came home, Dad was so proud that the house was what he thought was so clean. When she lifted up the rug, the dirt was swept underneath it.

One of the most important things our parents taught us was honesty. And to respect people and animals and to take care of what was yours. Family is important, and to take pride in what you are doing. I was expected to do well in school and to carry on and get a higher education, which I did. I attended the University of Calgary and graduated with a Bachelor of Arts degree, majoring in English with a minor in history. Whatever we decided to do, we were just expected to do our best. You know, they never said you should do this or you should do that. Whatever I decided upon, if I was happy with it and I was enjoying it, then I think that was fine with them. I think they thought my brother would carry on with the ranch because he was always interested in it. I think it was just a given.

After university I was a swimming instructor and lifeguard, and I worked for a while at *Alberta Beef* magazine doing advertising work, but I have always lived in the country, except while I was in university. If I had been a male, my choices in life would definitely have been different. There are certain things you can and cannot do – it's the gender issue I guess. If you are a woman you are expected to do certain things, although a lot of ranch women I know get right in there and do what the men are doing. Sometimes out of necessity and sometimes because they like to do that.

I'm starting to work outside a bit more now that the kids are older. Paul and I both feel it's important to be very involved in their lives. Sometimes it's a necessity for women to work away from the home to supplement their income. Often there's just not enough income from farming or ranching alone. I feel very fortunate to be able to stay at home with my kids while they are little and growing up. We both wanted kids. Of course, there are things you can't do when the kids are small, but I wouldn't ever say that was a compromise.

I want both kids to be well educated, and as far as being involved in the farm, I hope it will always be part of their lives. If not, that's fine, but I think an education is important, and they can decide if they want to pursue another career or come back and work on the farm or ranch.

One of the biggest changes is, I think, the personal computer. Brady started on it when he was about three, and I think he can work it better than his dad. That is a huge difference from when I grew up, and I think they are expected to know more before they even start grade one. I don't know if it is harder, but I think everything is changing and growing so fast. They are expected to keep up – there are a lot of expectations on these kids. Some of these changes are for the better, but sometimes I think everything seems to be so fast-paced now. I just would like them to take some time to smell the roses and to enjoy themselves.

I think a ranch is the best place to raise kids. You have privacy, the kids grow up and they are very grounded. They learn about the animals and how to have respect for animals. They learn about the crops and how to grow them. They learn that food doesn't come just from the grocery store, it comes from the crops, and I think that's important. It just gives them a very, very good background for anything they want to do. If they get a puppy or some chickens, they have to be responsible.

I volunteer at the school whenever I can fit it in, and I was treasurer of the Kathyrn ECS kindergarten for one year. I also took part in the Chinook Hereford Belles and I manned the booth at the Calgary Bull Sale, where we served coffee and cookies and it was a lot of fun. I think women have a lot to offer in these ranch-related organizations. Sometimes women will think of something that men don't see because of their different perspective, and women have some really great ideas. It's a matter of what they are interested in and if they have time and want to put their energies into something that is important to them – I think it is a boon to the organization to have women involved.

My life is not so demanding as Mom's and Grandma's were. You hear stories about women who had cloth diapers and they had to wash them and hang them out. Their daily stuff like that was a lot harder because we have automatic washers and dryers and dishwashers. I think the changes have been for the better.

It's hard to predict what life will be like for ranch women in the future, there's so much going on. Everything is so close to us that things that happen halfway across the world can have a significant impact on our kid's lives, way more than twenty-five or fifty years ago. I think maybe there will be fewer ranches, but I'm hoping it will be something my kids can do if they wish to.

There is nothing I can think of specifically that I would want to change about my life so far. Everyone makes mistakes and does things they wish they hadn't done, but in all honesty, if you don't make those mistakes, you can't learn from them. No one is perfect, so I think you learn from the mistakes you make, and just move on and make the most of life.

RUTH HUNT IN HER KITCHEN

RUTH HUNT

I was going to be a lady rancher and that was all there was to it.

Ruth Hunt grew up on a ranch at Endiang, Alberta. Her maternal grandfather named the town after a place he had in Muskoka, Ontario. Her father was an early settler at Endiang, and at one time their ranch was the largest in the area. In 1935 she married Wilf Hunt and they started out on their own in the "dirty thirties." In 1946 Wilf, Ruth, and their oldest daughter, Kathleen, relocated to their present ranch, eleven miles west of Airdrie, where they had a second daughter, Linda. Now a widow, Ruth, age eighty-three, lives in her own home and Linda and her family live next door and operate the ranch.

MOTHER DIDN'T HAVE much to do about running the ranch – she was definitely a mother and a housekeeper. She was always at Daddy's side and Daddy made all the decisions. Mother was in charge of the house and she always had help in the house. It was a big house – five bedrooms, and the hired men slept in a big room upstairs. She was very busy. She always baked big batches of bread, fifteen loaves two or three times a week. There was always ten or twelve at the table when I was growing up. Mother always had white damask tablecloths on the dining table – twelve feet long – two changes a week. My one memory of those is hanging them outside in winter and bringing them back into the house, frozen stiff, and then ironing them – two per week. My mother was always very well organized. She did all the patching and mending and darning for the family, Daddy and four brothers and me, and we had one man that was with us from the time he was eighteen until he was seventy. Mom used to mend for him too – he was like one of the family. And she embroidered and read, things like that.

We had electricity, a plant with our own generator. We didn't have the use of a lot of extra lights or anything. We just used lamps in the barn and we had coal oil lamps upstairs.

I was always a tomboy, always outside. I was second oldest with four brothers. We were all very active on the ranch – as soon as we were old enough and could work we all had chores. Our parents expected us to be honest, hardworking, fair, I guess that's about it. The younger ones were probably babied a bit more because my parents were older and there was more time, but they treated us all the same whether it was me or my brothers. No difference that way. I don't know what my parents thought, but I always said I was going to be a lady rancher and that was all there was to it. Never thought of anything else.

When we were kids growing up Dad and Mom used to go to Victoria, B.C. to visit Dad's brother who had originally homesteaded with him. He got kicked by a horse and spent the rest of his life in a wheelchair – that was Wilf's dad – and when he died, Wilf came out to the ranch to work because there was no work to be found anywhere else and he didn't have the education to go anywhere because he didn't ever have the opportunity.

Dad and Mom liked to dance – they wouldn't get to many at the ranch, but when we were at the coast they would go quite often. So we lived with Daddy's brother a lot when we were kids, and then afterwards Daddy bought a house there and he would leave Mom there with us kids for the winter, so we were going to school out there. I missed a lot of things because we sort of got mixed up between there and here. It was about grades three to six that we spent the winters in Victoria. I didn't enjoy going to school – I was too shy and everything. It was hard, but Mom and Dad wanted to get us out to show us there were other things to be had than where we lived. They just felt that we should have a chance to see more of life.

I went to a little school called Hunt Lake, about a mile and a half from home. We walked or rode or drove a team. We just learned the three R's, and the normal things. Going to school in those days out in the boondocks, there wasn't much chance of anything different. There weren't more than ten or twelve of us, one in each grade. I didn't go very far in school, I didn't finish grade eleven, and my brothers, they did finish their high school, and they went to Olds Agricultural College. I went to Agriculture College for two months and then I couldn't hack it, I was so homesick. We grew up living a sort of isolated life, and it just goes with you. I guess. If we hadn't been so isolated I probably would have gone farther in school, but I didn't try very hard. Our main time was taken up with work and we were pretty well tired when we got done with a day. It started early and it went late. I always milked and all those kind of things. Worked in the hayfield or whatever came up. I tried the piano but I didn't go very far. We had to drive six miles with a horse and buggy to get our piano lessons, so we didn't go at it very hard. I wasn't very good at it. My brother was better.

I can't say things would have been easier for me if I had been a boy. I guess males are stronger, but that didn't deter me in any way when I was growing up or after I got married. I did just the general things, whatever came up. When we were kids, it seemed to be my job milking the cows, riding – of course, my brothers had to ride too. And I worked in the hayfield too. I didn't ever have to stook and do that heavy harvest work – I didn't have to shovel and things like that, but I cleaned out the barn and looked after the animals. After I was married it was the same. Housework came second with me, sometimes third or fourth. And I just loved animals, just working with animals. I never was a cowgirl, I didn't profess to be a bronco buster of any kind. I never did own any fancy riding equipment or anything like that. I rode in a man's saddle – I never thought about things like that.

I used to harness my own horses and all that sort of thing. Nobody did that for me. Harness the team for haying. I worked with tractors here, but it was all horses when I was growing up. I did less riding after I was married because we didn't have the cattle for a long time when we were still out at Endiang. We moved here in '46 with the idea that we wanted to get into cattle, not just a small farm, and there was no pasture available where we were and we wanted to have more grass. We just lived for more cattle. As long as we were able to do it, we just worked hard together to acquire our ranch. We didn't have any help of any kind – it was just right from the very beginning in an old shack with no help.

When Wilf and I were married, I was seventeen. Our gift from one old bachelor was nine pullets. From my aunt we got a hen and sixteen chicks, and from another neighbour we got a basket of parsnips. It was all very much appreciated. We worked hard. And Daddy's motto always was "If you've got to borrow money, don't ever spend it on anything that isn't going to make a buck." So we were very poor, but so were all our neighbours. We lived in this old shack that was cold as the dickens in the winter, but we didn't think of it as being a hardship. It didn't hurt us at all. We were young.

We didn't have very many cattle. We didn't have a darn thing, you might say. Some of the land had gone back for taxes and relief feed out there at Endiang, and the farmers had left the land. Wilf and I were fortunate to get a half, and then we got another half, and then we got another half. In ten years we paid for all of that, as well as improved the fences and dug a well and government water holes and left it in good shape and came down here on a shoestring.

I was almost twenty when Kay was born in 1937, and I guess the first time I was pregnant, it was a matter of keeping the place warm. We had to pack all our water and pump it and all that sort of thing. We had coal and wood, mostly wood stoves. There wasn't too much work for Wilf to do because we didn't have the cattle. We didn't have anything. He worked the first few years for neighbours, and in return they did our land work, and things like that. We had no help. You couldn't borrow money in those days – we didn't have any collateral. Even if you did have some, you couldn't borrow – the banks weren't loaning money in those days. When we had Kay it was in the dirty thirties, and Wilf said, "That's it!" It was pretty hard. It's expensive now, too, but it wasn't only expensive – it was impossible. By the time Linda was born in March 1948, we had been at Airdrie a couple years and times were a bit better by then.

When Kay was born, we were twenty-eight miles from Hanna, and the only transportation we had was the train three times a week, so we had to go down on the train. My mother went with me and stayed with me off and on, because I waited in Hanna for a couple weeks. We were snowed in and we were afraid not to get down there. We just stayed in a boarding house for a little while with a woman who eventually got the measles and I had to move out. We ended up in the hotel for a few nights. Wilf couldn't get down to see me and I was in the hospital for quite a while. In those days they kept you in for a long time.

With Linda it was the same thing. We farmers around here had each given Vic Watson $100 to plough out the roads because we were snowed in, and we figured I better get out the next morning while the roads were open. We got a blizzard at two in the morning and it snowed us right in again. We left home here about 8:00 a.m. with a team and sleigh and our truck. The team and sleigh with the man in it were supposed to be digging out for the truck, but we finally gave up the truck and I rode in the sleigh and I got up to Glen Morison's place about 3:30 p.m. [a distance of about three miles]. It was so deep that they even had to dig the horses out. And then of course it was an awful thing to do to drop in on Morisons in that state, but Glen said, "Would you go in a plane, Ruth?" I said, "Yes, certainly" so he phoned up Eustice Bowhay, and he came out and picked me up. He landed right outside Glen's house there somewhere, and then in Calgary it was pretty rough because the airport was closed down and the taxis weren't running so we had quite a time. Mom and Dad had a house in Calgary at that time, so that's where I went for two or three days before she was born.

Wilf and I always made decisions together, on everything. He wasn't much of a housekeeper. If I wasn't here he didn't do much to get a meal for himself. It was just bread and jam or something. No, he didn't ever help in the house; I couldn't say that he did. The children were always right with us. Kay always had her 4-H calf and her horse. I can't say that she milked cows or anything like that. She was busy riding to school, quite a distance. When Linda and Kay were little, I still had to do work in the fields. It was pretty hard sometimes to know what to do with them. As long as Wilf had the horses and a binder he did it himself, but when he got the tractor it had to have two people. I started out by leaving Kay at the end of the row, and she played while I went up and down the row. Both my kids could pretty well entertain themselves. We didn't have any TV – they entertained themselves. So they went to the field with me if I had to go. When we were building this house in 1952, Linda was four, and the old fellow who was building the house – we didn't have a carpenter – she was happy with him when I had to go to the field.

The hours I worked outside varied with the seasons and the amount of help we had, and sometimes the help wasn't that great. And inside varied too, darn few [hours inside] at times. We always tried to eat square meals so I always had to see to that. So I would go out in the morning to the hayfield with the men and then I'd rush home to get lunch. At one time I had a bunch of chickens and I had to get the eggs while I was at home and rush back, and I usually ate my lunch while I was getting the others.

My youngest daughter, Linda, is married to Greg Sanden, and they have three children. The boy is a mechanic, which sure helps us out a lot. But then Greg is that way inclined too. The girls work but they all live here and they help do things like when we have to get the cattle in. Last weekend we had two big days, got all the cattle in and vaccinated them and things like that. Greg and Linda and I make the decisions. We are incorporated, and we work together at it. I hope I never have to retire from the ranch. It's hard not to be involved, that's my life. I keep the books, and we talk things over between us and I pretty well know what's going on around us all the time. It's actually become easier because we don't have to worry so much about where the next buck's coming from. We finally got the mortgage paid off, we don't have that over our head, so it's got to be easier. I have arthritis in my hip. I don't think that all the years working in the dust has helped my asthma, but that's never slowed me down much. I'm glad that Linda's here now, and her family, because it sure made a lot different life for me than having to get a hired man to do the work. They've been here and I've watched their kids grow up, it's all part of life.

Ranch women today have a lot more opportunity to get out and get involved with things than we ever did. In those days it was pretty hard for a woman to leave home. Travel by horse over bumpy roads, no cars, and they were pretty well stuck. I think that women should be involved in ranch-related organizations. I think it's a good mix. They have a lot to offer, and I think, why not? It's not necessarily a man-made business, is it? This ranch life, it's together, I think. Women have to know more about business these days. It's more than being able to ride or drive a tractor – you have to have business skills. Women have to be able to run a home as well as knowing what's going on outside.

To young ranchers starting out today, I would say, don't expect so darn much. Maybe my ideas are ruled by what I went through myself. I have no regrets about the way I grew up – I was always happy and I think my kids were the same. Lots of times they didn't have everything they wanted, but Wilf and I had a really happy married life and I was always happy when I was a kid. Even in my widow years, I'm fortunate. I wouldn't change anything. I'm quite happy. These days I think kids start out and expect too much. Look at the showers they get, and they sort of arrange to get at the wedding what they didn't get at the shower. They start out with everything, and I don't think that's good. I'm not saying that poverty is good for anybody, but I didn't ever consider myself in poverty. I know lots of people that have it tough, but I think they sometimes bring it on themselves – but that's the way life is. Tough.

My grandchildren are capable, but I don't know if they want something like ranching. It's hard these days. If you were trying to buy into one, I think it would be impossible. But if they inherit it, they've got to be pretty well dedicated, I think. I don't know as any of my grandkids will.

Ranching should be all right, it's a good life, but it's getting harder. This close to Calgary there are too many acreages. If you could get to a place like where I was born, it's quieter and land's cheaper. Ranching is just as good as any business. You have to like it and you have to have some kind of training. It's long hours lots of times with no holidays. You have to be willing to work.

CHERYL MORISON AT THE COMPUTER

CHERYL MORISON

The ranch wife has to be a partner with her husband.

Cheryl (Jones) Morison, age thirty-six, was raised the fourth generation on a purebred Hereford ranch east of Balzac, Alberta, and later, northwest of Calgary. She is married to Rod Morison and they live ten miles west of Airdrie with their two sons, Harley, eleven, and Carter, eight. They run a feedlot and a commercial cow-calf operation, as well as raising purebred Angus/Simmental cattle. Cheryl and Catherine Chalack are sisters.

I ATTENDED SCHOOL at Kathyrn and Aidrie, and after high school I went to Olds College and took livestock production for two years and got a diploma in that. I worked at home for a year and then I went to SAIT and got a certificate in professional cooking. I was twenty-four when I got married.

If I had been a male, I think it would have narrowed down my choices of occupation. I would probably be doing what my husband is doing instead of what I am doing because I am the mother.

I did more work outside on the ranch before I had children – then since having children, I do more work inside. I am not judgmental of people and what they choose to do, but speaking for myself, I am a stay-at-home mom. I don't see how you can spend all day doing a job off the ranch and still come home and be good at your commitments at home. Being a rancher's wife, I don't see how anyone could have time for both. I usually help with the horses, but if they are loading cattle I help, or if they are moving cattle I ride, but on a regular basis I don't help except with the horses.

Rod's parents and grandparents live very close and run the operation with him. Over the last four years Rod has taken over the management of the feedlot because this is something that has to be done spur of the moment, like decisions on grain buying and when to sell the cattle. You have to know the markets and use the futures. Whoever is making those decisions just doesn't have time to consult with everyone else before a decision is made. He consults with his father and grandfather as far as the management of the commercial cattle and the farming go. He is in charge of the purebreds, although I do the books and I also have some input when it comes to buying bulls. I also do the financial records for the farming, the cattle, and the feedlot, but Rod does the feeding records. Based on the experience I'm having now, I wish I had taken more business schooling in agriculture. Not bookkeeping particularly, just more management and financial planning. Also, some human relations training for dealing with employees would be useful.

It is sometimes difficult living so close to the relatives. Everybody needs to have an understanding of when to give each other some space, but I think it is good that Rod and his father and grandfather get along well enough to be able to work together.

I saw my grandmother in the house all the time, and that's pretty much where she was supposed to be, according to tradition. Because of modern conveniences, my mother was free to be outside and enjoyed taking part in all the outdoor activities and work with the horses and cattle. She also did all the yard work and gardening. If I wanted to be outside and work beside my husband, I could do that too, I guess. But I choose to keep track of the financial aspect of things, so I have less time to be outside because I do more office work. I hardly spend any time on housework and cooking. The big chore now is dealing with the financial end of things and keeping the books up to date. Along with technological advances in our society come extra burdens of work that you have to do. We have all these new so-called conveniences such as e-mail, Internet, etc., and there is so much information out there that we have to spend more time absorbing all that. Sometimes I wonder if it is just creating more work.

We involve the boys as much as possible. They love to help when we are processing feedlot cattle. It's a hydraulic chute and both boys are able to operate it. They usually have chores in the summertime, and this year Harley is in 4-H beef so he has winter chores as well. It's sort of a mixed feeling whether I want them to ranch or not. I would like them to be able to appreciate ranching and agriculture, but it is also important that they know how difficult it is and the reality of the economics of it. It is not an easy go. Mind you, they are the fourth generation of Morisons on this place, and because of that they may have an advantage over newer operations. I would like the boys to ranch if that's what they decide to do, but there are many, many other options open for them.

Whatever my sons decide to do, I hope that they will choose a wife that they love and who will be partners with them. The ranch wife has to be a partner with her husband. They can't compete with each other for time and energy because to make a go of it, you have to do it together. You have to make decisions about your ranch and your family together, and you have to be able to compromise one way or the other.

I am a member of several different ranch-related associations, but I don't have time at the moment to take an active part in them. I do think that women play an important part in these organizations, especially now with all the new technology.

If you are in ranching and agriculture for financial gain, if you are to survive, you have to be diverse. You have to go with advancement and technology. Ranchers have to educate themselves. They can't just stay out on the back forty and not realize what's going on with governments, world politics, and the whole picture. Agriculture is one of those things where you have to know the big picture to succeed in it.

From my point of view, being on a ranch, in our area that is, we have the best of both rural and urban worlds. We are half an hour from Calgary, we are ten minutes from Airdrie. There is every advantage and opportunity for the kids that you could think of, as well as proximity to excellent medical care. Except for the stress of trying to make a living in agriculture, I think it is a great place to raise a family.

CATHERINE CHALACK

If you have family support and if you have to contribute to maintain your ranching lifestyle, then it's okay for a mother to work off the ranch, but I don't think strangers outside your family circle are a good option for your children.

Catherine grew up among the fourth generation on Jones Hereford Ranch, six miles east of Balzac, Alberta. In 1978, when she was fifteen, the family formed a new entity, Jones Circle V Ranch, and moved to another part of the ranch, fourteen miles west of Balzac. The family raised purebred Horned Herefords. Catherine and her husband Tim Chalack live a few miles west of Carstairs, Alberta with their three children, Terra, ten, Riley, nine, and Kylina, seven. They raise purebred Black Simmental cattle, purebred Welsh ponies, and hunter ponies. Cheryl Morison is Catherine's sister.

I WENT TO KATHYRN SCHOOL, and then to high school in Airdrie. After that I just had freedom of choice and did what I wanted to do. I initially got a horsemastership diploma from the Pacific Horse Centre in Sacramento, California. Then I got a Bachelor of Education degree from the University of Calgary.

If I had been a boy, I probably would not have pursued horses as part of my career, especially not in the English

CATHERINE CHALACK DOING WHAT SHE LOVES

discipline – I probably would have stayed in the Western discipline. Or I probably would have felt pressure to remain on the ranch and work more with cattle and equipment.

I was married in 1988. I was twenty-five. At that time Tim was a dairy farmer at Cochrane, Alberta. Then a few years later we changed direction and have ended up here west of Carstairs with beef cattle.

Since my marriage I have been employed as a riding instructor and a part-time horse trainer, and I also taught in the schools at Cochrane. For the last two years, since my youngest child started school, I have been teaching equine science at Olds Agricultural College. It varies from five to ten hours a day, depending on what we're doing. That is an eight-month job from September to April. And then I'm home for four months, so my job complements what we're doing here at home.

If you have family support, and if you have to contribute to maintaining your ranching lifestyle, then it's okay for a mother to work off the ranch I don't think strangers outside your family circle are a good option for looking after your children. My children were never in day care. Tim's mom and my mom looked after the babies when I was at work.

Whenever I am here, there is always something to be done outside. I do everything from feeding the cattle, helping the farrier, helping with branding, vaccination, running to the vet, hoof trimming, halter-breaking bulls, pretty much everything in the daily routine. We have twenty horses here – we run a breeding operation. Some of them are strictly for the ranch, but we also have purebred Welsh ponies and hunter ponies – which are Welsh crossed with Thoroughbred or Quarter Horse.

Tim's quite good. He takes part in looking after the children. He makes sure that the horses are fed too, as well as the cattle, when I am working off the ranch. He's home at night if

I have meetings or when the children come home from school. If I am not home he makes sure that he's here.

If I didn't provide a supplemental income we would have to scale down our diversity immensely – we would likely have to get rid of the horse end of our operation completely, and probably have to focus on a very narrow scale with the cattle. We would probably have to do without a lot of extras – for instance, we spend a lot of time having the kids involved in extracurricular activities. One thing that my job offers is also some very secure benefits such as extended medical, dental, life insurance packages that are tremendous, and a really good disability insurance. Further, should my children choose to go to Olds College, I can get them some excellent rates on their tuition fees. The other disadvantage we would feel if I didn't have this job would be that our standard of living would not be as high.

Tim makes the majority of the decisions for the cattle and I do for the horses. As far as major family decisions like what direction we are taking, we make those together. The children currently own some of the cattle, and they take part in doing some of the chores and they have to help feed and help move electric fence and open and close gates. They are in Pony Club now and they will be in 4-H next year, so they will be more involved with the livestock. They like to come out and talk about selection for breeding too.

I don't really know if life is easier for women of my generation than it was for my mother or grandmother. I think it is different. Because of the opportunities available, it's easy for a woman to pursue a different career now. I don't think it's necessarily more difficult or easier as a general rule. I think it's an individual thing. I think if you are a rancher in an isolated area, you have to depend on your own household. But in our area, our family and extended families are large and we have probably a couple hundred family members within a forty-mile radius. I think it's a lot easier now than when people were first homesteading here. Our generation has a lot more support from families than even our parents' generation.

I think women are more empowered to be independent. I think they choose to educate themselves and give themselves something to fall back on. There is tremendous pressure to make sure that, even though the male in the household is the major breadwinner, the woman has a good education and a good background in work experience so that she is able to sustain the family in the event of a disaster, be it financial or physical.

I am not involved in ranch-related organizations in a political sense, but we are members of cattle and horse associations. I haven't been actually involved in anything bovine except the classroom agriculture program. I am an active inspector with the Canadian Warmblood Horse Breeders Association. That is, I serve on the committee that decides which horses are acceptable for breeding. I definitely think that women should be involved in these associations if that's their chosen career. Anyone who has a vested interest in these things should be involved.

As far as the future of ranching goes, we are going to be in trouble with the amount of land area that's available for producers. Government needs to be strong in lobbying for us in the future. I'm not saying that they need to throw a bunch of money at us, but I think ranchers should be considered a renewable, valuable resource in this province and this country. I don't think they are getting the credit that they need and deserve. And I don't think they're getting the breaks that they need and deserve. I'm not sure how to go about fixing this situation.

Ranch women in the future are going to have to be very supportive in making decisions with their partners. If they own the ranches and cattle themselves, they are going to have to have a voice, but the reality is, if you have a family and kids you have got to look after them as well. They are your priority and you can only do so much. It is important that women continue to write letters and continue to be vocal about the issues at hand and not to let those things die away. If you don't say anything, then you just end up being forgotten.

Ranch women are going to have to get very computerized very quickly, and I think they are going to have to have a lot of background in business because ranching is no longer just simply agriculture or being a good worker, having a good work ethic. It is about business, and that is one of the things I see that people are not good at – that is, the business side of ranching. The best thing we can do for our daughters is make sure that they understand business and have a good grasp on technology and how that will assist ranching in the future. It's not about having the biggest and newest equipment; it's about making good choices.

One thing I might change if I could is maybe I would have had my children earlier because now I can see the value of having children at a younger age, and I would encourage women to have their children younger if possible. It seems to be harder now to find a partner who is reliable. It seems that there are not too many people out there now who want to make a permanent commitment to marriage. They feel that it's just something to do and they can get out of it when they want to. There has been a huge attitude change in the last generation, and the attitude now seems to be "when the going gets tough I'll just throw this out the window and start again," and they don't think about the consequences. Right now the trend seems to be that everything is disposable. People like to buy things that are disposable and long-term commitments are hard to come by.

I wouldn't discourage my kids from ranching. It's a good lifestyle. But I would encourage them to pursue some education in that area before they made that choice. I think a ranch is a good place to raise children. I see from being a teacher and from just being friends with families in the city and small towns, just having this huge park area for the children to go out and run and feel safe and play, use their imagination and not be enclosed or confined, is a great advantage over kids living in the city who don't have that opportunity. One thing that ranch life does do for you is it gives you good solid roots, a good foundation, and the value of commitment. That's one thing we can do for our kids, give them the foundation and the roots and let them know that the onus is on them for their life. They make or break their life the way they want it to be. Something that ranch children learn at an early age, for the most part, is responsibility for themselves and to their world. I think they have a little more appreciation of the value of life and death and what it means to earn something and to try and hold on to it. So in those terms I would encourage my children to follow this way of life. It probably is not going to be easy.

LETA WISE TRAILING CATTLE

LETA WISE

There was nothing else I really wanted to do other than run this ranch. I just love the clouds and the sky and the prairie sunrise and sunset, and the silence.

Leta, daughter of E. J. C. and Helena (Loewen) Boake, grew up on the huge Boake family farm just east of Acme, Alberta. There were eleven children in the family, eight boys and three girls. They raised purebred Tamworth hogs and Jersey cows, and at one time had the largest herd of purebred Shorthorn beef cattle in Canada. In 1926 Mr. Boake bought a second ranch east of Irricana for summer pasture. Upon the death of her father, Leta inherited that ranch at the age of eighteen. She married Berwyn Wise in 1954 when she was twenty-one. Leta and Berwyn raised a family of four daughters and were in the purebred Shorthorn business, switching to Maine Anjou and Limousin cattle during the 1970s. Widowed since 1996, Leta lives with two of her daughters in the house she and Berwyn built, and together they operate Wise Maine Anjou Ranch.

I GREW UP IN A BIG FAMILY and I always felt like there were three different families because there was four older ones, and then three, and then there was us four young ones.

And of course the older ones always thought us four younger ones got the best of the deal, and maybe we did, but that wasn't our fault where we came into this. Anyway, everybody had to do their own part, and of course, being the youngest girl, I guess I got to do pretty well what I wanted to, like ride horses all day and work with the cattle. It was a big purebred Shorthorn herd and we sold a lot of bulls and females to the United States. That was quite a big market even then, as it is now. Also everything else went on there, pigs and sheep and chickens. We had to be self-sufficient. I could spend a lot of time just riding around. We didn't have a lot of horses but the ones we had were good. I always thought we should have more, but horses were starting to be a thing of the past when I got older. We still did threshing, that was a great time of the year in the fall. I had to get all the horses fed when they came into the barn from the field at night, and actually this year has been the kind of fall that just reminds me of the old threshing days. Day after day of warm, beautiful, clear weather. It was always so cool at the end of a day of threshing. To be a kid – have the lights on in the barn, horses fed – it would be very dark out. To hear the jangling of the tug chains and the harness on the teams – horses hooves clopping at a trot as they came closer and closer. And we'll never hear those sounds again.

That would have been in the early forties. At this time of year there wouldn't be a lot of chores to do and they'd be done by the time the men came home from the field – sort of had to be done in daylight – but I never milked the milk cows. Why I don't know, but I never did milk. Must have been aggravating for those brothers of mine.

A typical day in harvest time would be breakfast at 7:00 to 7:30, lunch in the field at 10:00, dinner 12:00 to 12:30, lunch in the field at 4:00, supper at 7:00 or whenever they quit for the night. In the earlier years I can just remember there was a cookhouse and a cook for the threshing crews of eight to ten men.

It was in '45 the first time I ever showed at the Calgary Bull Sale. Of course, all this time it would have been my brothers and my dad that were looking after the cattle, and they were the ones that took the bulls to the Calgary Bull Sale. Calgary Bull Sale was the big thing. Until I was eighteen I could show in the Baby Beef. It was a totally different ball game than it is today. Girls just weren't quite accepted. I don't know about women, but they thought that girls should be home. After dark they better be in, out of sight. So it wasn't always so comfortable being there, you know, but I was. And some people thought it was maybe all right, and maybe some people didn't. I was all right with it, but still when you think about it, things were really different. Nowadays all the women are there with their families, everybody is there and it's one big family, but things weren't like that in the old days. 'Course a lot of things aren't like they were in the old days.

What my mother did was work. That was cooking and washing and raising the babies, and I just have to shake my head now that anybody had that much work. There was no running water and I always used to cuss my brothers because they wouldn't go and haul the water. There were eleven of us when we were all there, and then there might be a couple hired men besides, so that was quite a few people around the table. A lot of meat and potatoes, a vegetable, and maybe a cake or pie. Just what we grew ourselves. When Mother had time to sit down and take a break, she'd pick up her knitting. There was a lot of that to do too. The knitting was a form of relaxation. It came so easily and it was nothing to knit up a pair of wool socks or mitts. I remember one year during the war years that Mom knit twenty-two sweaters for the Army and the Air Force. She was really proud of that. The wool was given and the women produced the sweaters. I guess that this would have been besides the regular wool garments that were needed for the family. I think my mom had it tough, and the more you know about it … you think that, God, day after day it was diapers and dishes and none of the modern conveniences. She didn't have help in the house very often. There was Ann and Mary, my two sisters, and I helped, but I couldn't wait for those brothers to get married and get out of there and let somebody else do their laundry. Mary was the first one to leave and go become a nurse, and then Ann, and then after that I had to do the laundry.

We didn't actually have that big a house. There was a bunkhouse for the hired men, and my brothers lived in the bunkhouse too as they got older. The hired men and my brothers would maybe get a little wood or water sometimes, but they never helped with any of the work in the house.

My father was the big kahoona, the boss. My father made all the decisions – there was no togetherness. He would do all the fixing of the machinery. He would have done a lot of work gardening here at the ranch, and of course, he planted all these trees. He and my brothers. This was his dream, having all these trees, like he thought someday they would be a crop to be harvested because he came from Ontario. There are a lot of elms and maples planted. We had a big garden, fruit trees and everything. He was down here a lot and he had the irrigation here, so that made things easy to grow. We are very thankful for all this.

All I saw was horses and cows. I know my sisters were going to be nurses and be on with something else, but not me. It was horses and cows from the beginning. We were all expected to work hard, be honest, no buggerin' around. The boys were expected to get their work done and then they could go out at night and goof off, but the girls better be at home. Didn't need any boyfriends, didn't need anybody coming in.

Dad never discussed any plans with us. Hell no, he never said anything, but since he left me this ranch, I suppose this is the way he meant it be. And the boys, I guess he just thought they'd work and run the farm. Their wives didn't always like that too good. He wasn't prepared for the boys growing up and having families. Everybody can't live in the same yard, but back then that was the way it was, and when you've got a big family, well… I don't actually know what Dad was thinking. He was only sixty-eight when he was killed in a car accident. Four brothers were married then. Things changed drastically after that and things got split up.

I went to school at a little country school, Delft School, until grade eight, and then we went to another country school, Simcoe, for a couple years, and then it was school buses to Acme, where I finished grade twelve. My dad always said you had to go to school; you had to learn, it was important to finish school.

If I had been a man, I suppose there would have been more choices, but I was willed this ranch, this place. I couldn't do anything about that, you know. The ones that complained about it were the ones that got most of the farm, and I don't know why they complained. Why should they? I have lived here ever since I was married when I was twenty-one. I married Berwyn Wise from west of Irricana, and he actually had nothing to do with beef cattle at that time because his family had a dairy, so I think it was a real challenge to get into something new and make that work. He learned a lot about judging cattle and feeding cattle and all the rest of it. Some of it he learned from me, but he learned a lot from other people and from reading. We pretty well made the decisions together, you know, about buying bulls and selling cattle – we did that together. When we were setting goals, I sorta let him lead the way. You know you must keep the men happy. You gotta let them think they're the boss. When the girls got older, we got into the horses a little bit and showing horses. It was new and a challenge, and to me that was the big thing.

I did all of the housework and as much of the chores as I needed to. I always did a lot of the riding. I did all the books that needed to be done, the financial records and the cattle records. I like to get that done. It's not that my husband didn't do anything – we did these things together because we very seldom had any hired help. And we started to have a family and I had to look after them, and I still did the other work I did before. Life was really simple in the beginning, really simple. Now, as I see it, the woman has to end up seeing that it's running because it's got to work. If there's a lot of squabbling going on, it's tough on the kids and everything's gotta be copacetic. I guess the right word is *peacemaker* the woman has to be the peacemaker.

I did proofreading at the local newspaper for a while. Back in about 1986 I had a job registering cattle with the Canadian Salers Association. I loved that. Both jobs were learning experiences that I really enjoyed. But I've always lived here on the ranch.

Berwyn helped in the house some, but not a lot. He could cook, he learned to do that at home because his mother wasn't healthy, but he didn't have to do a lot of that. He didn't do any laundry. He'd hold the babies once in a while, but they were good kids – I never had any problem with the babies. I nursed them all and that made it much easier.

I have four daughters. Debbie is forty-six, Deanna is forty-four, Della is forty-one, Dallas is thirty-two. Della is the only one married. She has two girls. Debbie was married but she is divorced now, and she lives in Colorado. Deanna and Dallas live here with me. I stayed home with my kids. I think mothers pretty well need to be there when they are little. I actually don't know why mothers have kids and then put them in day care and go to work – I really don't know why they do that. And then the kids grow up and the parents have a terrible relationship with them, and why did they even have them? And then the parents wonder what happened. Well, you weren't there!

But once the kids go to school, that's a different thing, if you can arrange to be home when they get home from school.

My babies were all born in the General Hospital in Calgary, but we were all born at Grandma Loewen's. She was a midwife. She learned this mid-wifery from Dr. Fowler and Dr. Elliot, and once her family grew up she converted her house into three maternity bedrooms and she had maternity women there all the time. She lived only three miles north of the farm at Acme, so we were all born there except my youngest two brothers.

Berwyn died in the spring of '96, and Deanna and Dallas and I run the ranch. It's all my responsibility really. Deanna's not quite at the point yet that she can take it all over. You've gotta have that financial net – you've gotta have an operating loan and you've gotta be big enough to be able to handle it. We confer with Debbie because she's still involved – she has some cows and I lease them from her so we need her input. Della and her family live close by so I see them quite often. Deanna looks after all the equipment and the field work and the cattle too, but not the riding. Dallas and I do the pasture work and the cattle after the calves are on the ground. Things change after they go to pasture – then Deanna goes a different direction and we look after the pasture. The girls could do what they wanted and they pretty much did. I'm very satisfied with everything they've accomplished. If they wanted to take some different schooling, I just paid for it and go ahead and do it.

Della assists the teacher at the Hutterite colony half-time, and she just loves teaching. Deanna has her photography business, which she needs to have, and Dallas has a job in a doctor's office three days a week and then she's home for four. Dallas also does Reiki and has made a profit from painting watercolours. Even though the cattle business is good, you still need this other cash flow. Berwyn and I got along all right without, but I suppose it's because everything is so much more expensive now.

I think my kids had a better life growing up than I did because of the atmosphere in which they lived. These kids were much freer. Nobody said you can't do this, you can't do that. Not that anybody said I couldn't do anything at home; it's what you know. In my home Dad was it, you know, and you toadied to how he thought about things. Don't upset the lion, that's absolutely right all down the line. Don't upset the lion. Keep it together.

For sure a ranch is a good place to raise children. My granddaughters come over and they just love it here. They come over when we work cattle, and right now they've learned the numbers, the weights, and they've learned to ride and they just love it here.

I'm not at retirement age yet and I have taken care of everything. It will be just turned over to my kids someday so they don't have to worry about that either. They know what's coming and they can plan. And right here where we live is very cheap to live, – taxes are low, and then having irrigation is very good, so they should be able to do all right here. But my health is good and I have no vices, no smoking, no drinking, and no men, and I'm right happy.

Although I was denied the college courses I wanted to take because I was a woman, I did have more choices than Mom. There are a lot more choices now, and I think it's much better. That does not mean we leave behind the tough decisions or the hard work, but it means we have the opportunities to follow our dreams. In cattle organizations there are more women involved, too. There might be men out front, but I think the women are right behind there prodding them because there's just women everywhere now. They play a much bigger role. I think it's good. I am on the board of the Stockmen's Memorial Foundation. We raise money to run the Bert Sheppard Library, a historical library for ranching in Western Canada. It has all the old brand books and all the old RCMP records, people's writings and papers and local history books.

To run a ranch on her own a woman has to be pretty into tough work, 'cause it's hard work. There's a lot of manual labour, and you've gotta be able to do all your books and the financial and business management. And then you've got to be able to travel and go to all these different functions where you learn what's going on. That's something we never had to do. It wasn't such a big thing, but now there's all sorts of courses to take and you need to do all those things. You just have to persevere, and it's not always easy. There's good years and some are not so good, you gotta be able to take all that in stride.

There was nothing else I really wanted to do other than run this ranch. At one time when I was just finishing school, Charlie Yule thought I should go to Washington State and he wrote a letter to them for me. I got this letter back saying they didn't actually have any courses in agriculture for girls, so maybe home economics would be better. So that would have made a big difference if I had gone to college. It probably would have changed my life a lot, but the choice wasn't there.

I think this ranch will always be home for my daughters because this is where they can come to live regardless if they have cattle on it or not. If they stay with the cattle, they'll be part of it unless they decide to do something entirely different. I think there's a future in ranching, I can't see it changing. I think people thought a long time ago that things would change, but they haven't much.

One reason why it's so good to ranch here is that you can get on a horse and go out and take a ride and check cattle or whatever, and just take away all pressures and come back a new person. And there is a certain wind that needs to be there too; it's not a hot wind, just this breeze. It's just exhilarating. I don't ride as much now but I walk. I feel very fortunate to have been able to live my dream, to see the calves, the foals, the grass, my daughters, and now my granddaughters, grow on the ranch I love. I just love the clouds and the sky and the prairie sunrise and sunset, and the silence.

EILEEN MCELROY CLAYTON

Let's face it, there have been women working on farms and ranches, unheralded, for generations. The biggest unpaid labour force in the country.

Eileen lives just east of Chestermere, which is on the eastern edge of Calgary. Eileen's father, Morton McElroy, recently passed away at the age of 89, so now she and her mother, Joyce McElroy, run the farming and ranching operation. She also assists her husband, Warren, with his cattle. Eileen homeschools her daughter Heather and son Kent. The McElroys also have some pasture northwest of Calgary, and both there and at the home place; they are threatened with Calgary's, and Chestermere's, urban sprawl. Eileen has one brother, Donald, who does not farm.

WHEN I WAS GROWING UP, I think Dad made most of the decisions about the farm and ranch, but for the family and everything else I think Mom and Dad jointly made the decisions. Mind you, Mom always had her own cattle, too. Dad said when they got married that a woman should always have her own source of income. It was fairly modern of him for a fairly traditional man. But I think he had heard of some lady having to ask her husband for money for a dress or something, and he thought that a woman shouldn't have to do that. And of course, my mom came from a ranch in eastern Alberta, near Empress. Part of their place was on the British Block, and they had to move off when they put the Block in there. So she was a ranch girl.

My mother's main work was the house and the garden. When I was growing up, there was a really big garden because there were hired men. In fact, I remember when I was still quite young, it was a huge potato plot and they were using an old workhorse in ploughing up. So there was a lot of weeding and stuff to do there. The cook usually would help. Not so much with the weeding but in the harvesting of it and things. But there's a lot of work just to running the cook. If you ask my mother about the hiring of the cook and the different cooks! I think there was a lot of relief when there weren't so many hired men and we didn't need a cook. And she was into volunteer things too. She worked on the Cancer Society and the history book, and she belonged to Women of Unifarm and quite a few things. She helps check for sickness in the feedlot and with calving. That's one place where I think women do have a slight edge – even in some of these bigger feedlots, many prefer girls to check the pens. I guess it's our gentle, nurturing side that can pick out the sick calves and treat them. I always was interested in the animals as far back as I can remember. And I did like working with the cattle and I loved to go up to the ranch with Dad in the summer and check cattle. I remember that from when I was really young. I used to think I never would get married – I just wanted to do my own thing basically.

I don't remember my parents ever saying, we want you to do this or not do that – other than the underlying expectation that you would work hard and do your best at whatever it was that you were going to do.

I went to grade six at Conrich. Then in grade seven I came over to Chestermere, where I finished senior high. And then I went to university, first in Calgary, and then I got my Bachelor of Science in agriculture from the University of Saskatchewan in 1981. I think at Conrich and Chestermere we got a good basic education in that we learned to read and write fairly well and to do basic math. Computers weren't an issue then, but I wonder if now they put too much emphasis on the extras that you can pick up because you need them later, and not enough on being literate and numerate.

I don't think being in the country affected our education. I guess it depends what you mean by a well-educated person. If you are just going to depend on schooling and never learn anything after that, well.... If you try to put in too many enrichments, I think you are sacrificing the basics.

After I got out of university, I worked at Canada Packers as a sales rep in the food services department, selling to hotels, restaurants, institutions. That was a very interesting experience. I got to see another facet of the industry, the end user sort of thing. And I got to see price lists and how they didn't correspond to the price of fat cattle. When I left Canada Packers to come home to the farm, I just worked on the farm and that's what I did.

When I got home, my cousins and Dad and my brother were doing the work, and hired men too, but I was at the bottom of the pecking order so I did get a lot of jobs that other people didn't want to do, but that was fine because that's sort of how it goes. You start at the bottom and work your way up. My cousin, I remember – I don't know whether he did it on purpose or whether he just had a feeling for the jobs I didn't like – but he gave me those jobs. Like climbing up to the top of grain bins, and I had a dreadful fear of heights, which I still do, but it's manageable, likely due to that. I did everything. Picking rocks, fencing, seeding, harvesting, working the cattle, feeding in winter, bedding. I didn't learn to weld because it seemed to me that there were plenty of people there who knew how to do it. And I didn't learn to run the air seeder, but I think I ran just about all the other machinery.

I had to make some compromises when the kids were little. I had this idea, and I know my mother was chuckling to herself and shaking her head. I had this idea that I would just sort of drop this child and in two or three weeks I would be back at it. I don't know why I thought that – it was pretty unrealistic. My labours took a long time, probably because I was older I think I was thirty-one when Heather was born. I remember thinking just before she was born that either I could die or the baby could be born, but I just couldn't keep on with this forever. Because I'd had viral meningitis the year before, they wouldn't give me a spinal block or anything. She was born in June, so I can't remember whether I went back to feeding cattle that winter or whether it was the next year. But for sure it was the next year, and then Mom would keep her while I was working. But when Kent was born, I'd still ride in the summer to the ranch and herd cattle and help with branding and all that. The kids are bigger now and they get called to do stuff if we are branding or working cattle because Warren doesn't have hired men. And this last March when we had the bad weather when we were calving – we don't have a heated shed, and the kids were really good. We'd bring the calves in and dump them in the hallway and they'd get the hairdryer and towels and dry them off.

As far as raising children on a ranch, I think it beats, by a lot, the alternative, that's for sure. There is a certain amount of reality check on a ranch, that's for certain. On a ranch, kids know pretty much what mom and dad are doing. You know where your daily bread is coming from and what it takes to get there, and so on. I just think it's healthy too – my goodness, little kids turned out in a backyard in town? There's just not enough room. It's a great environment for children. Of course, that is the way I was raised too. I guess I don't know anything about the way it would be to be raised in the city. I like to be outside and I guess I'm not much of a city person. It's just the peace and quiet, and you can do your manual labour and think. It's sort of intangible in a way. I guess I can say I don't have to get up and beautify myself before I go to work, and I don't need an extensive wardrobe.

I'd be happy if the kids ranch here when they grow up because it's something that has been in the family, but it's not something I would force on them. I want them to be involved in something they are passionate about, something where they feel they can make a contribution. But I wouldn't want it if it were not what they want to do. It just wouldn't work out. You have to go where your heart is.

I have been on the board of Western Stock Growers for about twelve or thirteen years. Then I have been treasurer for about eight years. I was the first woman to be on the board. Even before I was a director, I was around the Stock Growers quite a lot and I helped on one convention before I was on the board. The attitude was if you could get somebody to assist who was interested and would put some work into it, they would be happy to have you. I can't really speak for them because I don't know what they said when I wasn't around. But they seemed more than happy to have a warm body to help run the place. So everything was positive – I didn't have any negative comments or I didn't feel that I wasn't wanted. I got on there and they put me to work. After I had been on the board for some time, another ranch girl, Cherie Copithorne, was on the board for a while. Then a year or two ago a lady from out in the Hanna area came on board, Karen Gordon, so I have a compatriot.

I heard a few funny comments, like when I was expecting Heather, they said I was the first director to calve while I was a sitting director. I remember there was one fellow, he is passed away now, but when we were having our pictures taken he would say, "Here, stand here, it's a rose between two thorns." But I never got any put-down comments or anything like that. By the same token, they did not treat you with kid gloves like you couldn't do things. If there was a storage room to be cleaned out, you could lift things just as well as men. I think, generally speaking, that ranchers are more conservative and traditional than maybe some other occupations, just the nature of the business. But certainly they are not chauvinists – they are gentlemen, and actually in this day and age it is quite refreshing. It has been an absolutely wonderful experience. You know, a lot of work, sometimes some trials, but it has been a privilege really. You meet some wonderful people, and people who have given decades to the livestock industry. You just learn so much more than what you put in.

There are a lot of ladies, be they wives of directors or whatever, who have no official position and are not on the board per se, who do a lot of work, and you couldn't run a lot of the events without them. When we had our big centennial celebrations in '96, there were ladies who went and did research for the book that was written. They spent hours going through the files at the Glenbow. Archives. And also when you are on the board, whether it be a man or a woman, the partner has to facilitate that because someone has to cover for you when you are gone. If Warren wasn't on side, I wouldn't be able to do it because he and Mom and Dad have had to cover for me when I have been away at convention and other meetings. So you can't do any of these kinds of things unless you have your family on board.

Dad's mother was a city girl from Scotland, and she married my grandfather when she was thirty and my grandfather was forty-five. She was a very proper Victorian lady. My grandfather took her to the very first Calgary Stampede [1912] when they were courting. When the local ladies were organizing some kind of ladies' group, she asked her husband if she should go to the meeting, and he said she should stay home with her boys. Certainly in that era there were lots of ladies here who were active in lots of things, but not my grandmother.

All our domestic things are so much more automated now. Just the basics of meals, laundry, childcare, were so much more difficult. A lot of rancher's wives had a hired girl, but there was still such a tremendous amount of work. It wouldn't have left much time for anything else. I know my grandmother loved to sing. She was a concert musician – she played the piano and sang. She used to do that at church and some of the local functions. But I think it would have been a very uncommon situation for women in those days to be involved in any of the organizations such as the Stock Growers or any other lobby group. My goodness, if you go back far enough, they didn't even have the vote, so it seems unlikely that they would have belonged to any such organization. A single woman might have a little more leeway, but a married woman with children probably just wouldn't have time.

Even in the time that I have been on the Stock Growers, it seems like there's less and less fun and more and more serious challenges facing agriculture. Where urban people are concerned, they used to have some connection with agriculture, but now it is getting less so – they just don't understand. There are lots of misconceptions in the media. You know, you keep hearing about these factory farms and the urban people begin to think that's what farming is like. As far as women in agriculture, presumably as you go along you face fewer problems that come up because you're a woman, but I don't think I faced any problems because I was a female. Let's face it, there have been women working on farms and ranches, unheralded, for generations. The biggest unpaid labour force in the country. They have always been a vital part of the operation not just for domestic work but for the outside work as well. Or ladies with second jobs, such as teachers and nurses, who have brought the farms through rough economic times.

To run a ranch you need the skills that you would need to run any kind of a business, because it is a business. You need perseverance, you need optimism because bad years come, like the droughts we have had and the terrible spring last year. And maybe for a ranch and farm, specifically, you would need to have some liking for the values and the lifestyle.

In the early days I'm sure there must have been some ranch women who were pregnant at the time of their marriage, and they wouldn't be unique. It's been interesting to me, since I've been doing a lot of genealogy, to discover that there were a lot of very short gestation periods. It was much more common than I would have thought, seemed to be a lot of it.

If I had to do it over again, I probably would have put a little more oomph into some challenging things when I was younger. You don't realize when you're in your twenties how much energy you really have, and if you're going to move mountains and shape the world, you should do it then. But on the other hand, maybe you don't have the wisdom that you'd need either. I don't think I would have taken anything different in university. But you know, I would still get married, I would still have kids, even though I thought I wouldn't. I wouldn't want to live in the city and have a city career.

I guess if you are in a family operation, whether it be with parents or brothers or sisters, you have to be prepared to compromise, to sometimes step back, and hopefully the other people are too. Otherwise it just doesn't work. If it's something you really want to do, you may have to make some sacrifices. For instance, when calving season is on, you won't be able to go on a holiday or anything, or during haying season, you are at home. If it is something you want to do, you just have to stick with it. The rewards are there, – they are not always monetary, quite often not monetary. But there is, I think, a lot of satisfaction in being a food producer. I think you are more in tune with the big rhythms of life. You realize when bad weather comes along that there are things bigger than you that you just have to accept, no matter what. And you have to just do whatever is required to get through it. Whether it is a cold calving or the drought, these things are definitely hard on people, and I think a lot of farm and ranch families come to a point where they just can't go on. You meet those kinds of challenges, and if you're in a position where you have a loan, it's a pretty scary thing.

It's good to listen to what older people have to say. It has been my experience that they have a lot they can teach us if we will just listen. I know I've learned a lot from older people in the Stock Growers. And also from my dad, of course, who has lived through a depression and seen some pretty hard times.

I'm sure we can't stay here until we get old if we get really old, because this Chestermere town is closing in on one side and there are more and more acreages. I suspect we would probably try to sell this property for acreages, and then at that point, whether we move to another ranch or farm is the question. I can't see that I would ever want to retire into the city.

KIM TAYLOR LOOKING THROUGH HER CAMERA LENS
WHILE MOUNTED ON 'SHUFLY'

KIMBERLEY TAYLOR

Ranch women are probably the most versatile people on earth,
mainly because there really aren't many jobs they can't do,
if they put their mind to it.

Kimberley Taylor is the fourth oldest of eight Bakken children who grew up near Bracken, Saskatchewan, where her family always had cattle and horses. With camera in hand, Kim is dedicated to preserving the Western lifestyle on film, and the forty-eight-year-old, recently featured in *American Cowboy* magazine, is rapidly gaining recognition as one of the West's premier photographers. The images she captures are, in essence, a reflection of her own life.

OUT OF US KIDS, four girls, four boys, the girls were the ones that pursued careers with horses and cattle and ranching. In the summertime we rode horses and played cowboys and Indians. We swam in the little slough. We used to pack a big picnic lunch and go to the river hills on Sundays. Mom and Dad would roam the hills and look for artifacts, and we'd take tires and push them down the river hills and use up tons of energy.

We spent the whole winter tobogganing and skating. We used to roll up in a quilt under the yard light and sleep outside and count the stars, and look at the Big Dipper. When it was too cold out to be outside, we had card games and jigsaw puzzles.

I left home when I was fifteen, went to live with a family in Montana, and I graduated with grade twelve from Whitewater, Montana. It was hard for me to stay in school. I'd rather be outside. My sister had married an American and was working on the ranch where I went to live, so I had that connection. They probably ran about five hundred cattle at that time and had a big A.I. program, so he sent me to A.I. school.

When I came back to Canada, I worked in an arena looking after horses, in a Western store, and then at a community pasture at Central Butte, where I A.I.'d and rode. I wasn't a very good roper then, and we had to treat this one bull and he looked pretty quiet, so I thought, what the heck! So I got off and sprayed his eye with a bunch of pinkeye spray and got back on and came home and told the boss which bull I treated, and he's like, "Oh, my gosh – you did that?" I guess the good Lord must have been looking out for me.

The A.I. program that we had – we used to get up about three-thirty in the morning and we'd saddle up and ride until probably ten o'clock and then we'd come back in, usually have something to eat, sleep a few hours, and go back out about three-thirty, four o'clock and check again until dark. We pretty much did that day after day. When you're eighteen, you're kind of wanting to get out on Saturday nights sometimes.

I met my future husband at the pasture at Central Butte. We got married in July of 1975 and he was working for community pastures at that time. We lived in Red Deer for one year, where he was a brand inspector. Then we moved back and worked at the community pasture at Birsay for three years, and I learned to rope and heel calves there. Probably my favourite part of ranching is heeling calves in spring and cutting and sorting calves in the fall.

We moved to Kerrobert for three summers, Elrose for a year, Smiley one winter, and six years at Bracken. Jill was born at Eatonia in '79, Jeffrey at Elrose in '83. When I moved to Bracken, I had moved something like seventeen times in eleven years. I found I looked at the kids' lives and all our moves as chapters. I made my life into a book.

I had a great mother-in-law. She used to come down every spring and fall and do cooking and babysitting so I could get out and ride and help sort. I'd know ahead of time and have things frozen in the deep freeze and have most of the meals organized.

And then we got cattle of our own, and we'd have those to calve out in the spring, too.

We moved to Maple Creek in '89, and that's when things kind of really opened up for me in discovering my photography and starting to document the ranch life. I met Nick Demchenko, a saddlemaker, silversmith, and horsehair hitcher; the Murray girls, who are artists; Ken Wright, who is a saddlemaker; and numerous others. I was in the core of ranching country and it kind of started the creative juices flowing. I had no formal training or education in photography, and it started out basically as a hobby. I met Jan Daley, who at that time was doing a lot of photography, and she helped me. It just seemed one door opened after another for me when I started my photography.

I had heard of poetry gatherings, one down in Big Timber and one out in Pincher Creek, Alberta, and I talked to a few people in Maple Creek and said, "Hey, you know, Maple Creek should have one of these." I helped organize the Maple Creek Gathering for ten years until I moved to Alberta.

I worked in the auction mart at Maple Creek for a few years. Most of the time I was just at home to be with the kids. I remember explaining to my son one day – he'd come back from Agribition and he was talking about one of the kids that was with him – that his mom had given him so much money to spend while he was at Agribition. And I said, "Well, there's two ways that you can be a mom." I said, "His mom chose to be a career person and work out of the home and that allowed him to have a lot more money than we have, but I chose to be at home with you kids and be there while you are growing up." We drove a few miles and he never said anything. And all of a sudden he looked over and he says, "I think you made the right choice, Mom."

My business was becoming quite established in Maple Creek, and when I moved to Alberta that first year, it was quite a struggle. I was on my own. People didn't know me. I didn't really have a place. I moved around a lot and I took whatever photography jobs I could find. I felt quite helpless in a lot of ways, felt the only life I knew was ranch life. I started working with *Canadian Cowboy* magazine quite extensively, and I did a story on a rancher at Bragg Creek. When I came back to visit him, his health had failed quite a lot, so I offered just to come do chores if he needed any help.

Since I've moved to Bragg Creek and started working on the Connop Ranch, I've had probably the most responsibility as far as ranching goes that I've had during my whole life. I looked after all the calving and Jim was in the hospital from July to November, so I had the responsibility of looking after everything while he was in there. The first spring we had like five feet of snow in one fall and we were really immobilized the next morning. We were feeding and were just starting to calve and we had a set of twins lying on a big snow bank when I went out that morning. I got them in and I saved one but the other one didn't make it. We had to get a Cat out to make paths so we could get to the feed to feed the cows, and we fed in little ruts they made with the Cat. And you're out there from the time you get up in the morning until dark at night, just fighting the snow. We were trying to bring cattle in that were getting closer to calving, and it was so deep, there were places you couldn't even go with your horse.

The spring I came there, we had a bear come in and take down one of the yearlings. It killed the first one and mutilated the second one enough that we had to put it down. The first time I rode through the bush and saw all the claw marks on the trees, I thought, I've got to go back to Saskatchewan. But I understand their nature a little bit more now and know that I'm in their territory, they're not in mine.

Branding day – I usually cook for fifteen or twenty people, and I'm cooking before I go out in the morning. I like to be out at the corrals working, so I do my part out there and then come in as soon as the last calf is done and get the meal on. I was never afraid of work and I've never felt that you had to take a back seat or that you weren't accepted or that you were lesser because you were a woman. Most times if you could do the job, then you were just part of the crew.

The deal that I made originally was that I'd work part-time, but ranch life kind of dictates how much time I'm really going to spend there. I found a lot of days were just full days on the ranch and my business suffered – I was really wearing myself thin. So I renegotiated this last summer and there's another fellow now that helps at the ranch. My business has really picked up and come together, and I'm a lot freer to go out and do the shooting and travel and marketing now that I don't have all the responsibilities of feeding the cattle. I miss it though. It was as much a love and a passion as my photography, but my photography is what makes my living and so I had to make a choice – I think I bawled for two days after.

I've continued to become better with horses and now I've got a Doc o' Lena-bred gelding, and he's pretty "cowie" and willing, so it's a lot more fun when you go to sort. I don't enjoy riding a raw colt. I like to ride them after they've had about a month's riding on them, and then go from there, but I do enjoy doing a lot of the groundwork with the horses. It's amazing, their intelligence – as we learn, how much more the horse can learn. When I saddle up I want to know I've got bulls to move or cows to check or something to rope or sort or whatever. It's just such a satisfying feeling and I guess I like that versatility in my life, too, because one day I can head out in my boots and jeans to the barn and the next day I'm dressed up and headed downtown for a business meeting to market my work.

In 1996 I documented the Western Stock Growers hundredth-anniversary cattle drive. We trailed sixteen hundred head of Longhorns from Buffalo all the way down into Medicine Hat. There were twenty-three hundred people and I think one hundred and three wagons. When I found out where we were going, into the Suffield area north of Medicine Hat, I thought, why, in all of Alberta, would we pick that place, but as we went across there it gave me such a feeling for what our forefathers must have felt when they first came into this country, not knowing where the next water hole was. You could ride for miles and miles and not see a fence line or a power pole. It was just the raw prairie and rattlesnakes and rivers and whatever was out there. I guess I have always had the pioneer spirit in me and it's the lifestyle of those people that have the ranching roots that I want to preserve.

This year will be the tenth year that I have had my business – Slidin' U Photography – and I publish a day planner. I've done that for the past five years now, and I recently did a black and white wall calendar and I've done a company calendar that I sell to businesses. It's quite interesting now to turn my computer on in the morning and have emails from Holland, the UK, Germany, and all over the States, even New York. One of my goals when I started my photography was to show our way of life to people that couldn't geographically be here.

When I got the idea to do this day planner, I needed, like, sixteen thousand or something dollars. I went to three banks and none of them would lend me money because I didn't have the collateral and I didn't have any credit rating, so it was really discouraging.

This was one time when it really paid to go with your gut feeling. A friend called up one evening and I answered, and he's, like, "So, how are you tonight?" And I went, "Oh, I'm fine." And then I went, "No, you know what? – I'm not," I said. "I have this great idea, and three banks turned me down and the layout artist that I was going to use said she couldn't work the program that I needed so," I said, "I'm really frustrated." So he said, "Well, what are you doing?" So I explained to him and he said, "Well, maybe I could help you," and I'm, like, oh, I couldn't take help from, you know, a neighbour, kind of thing. So anyway, I thought about it for the weekend and I thought, you know, he's the kind of guy that wouldn't have offered if he didn't mean it. So I went back to him Monday and I took down what I was doing and showed him everything. And he said, "Well, I have always liked your work and believe in what you're doing, so I'll phone my banker and see if we can help you some way." So that Wednesday I sat in his banker's office and got the money I needed and away I went with my idea and I've never looked back since. In fact, I think I had the twelve thousand that I had to borrow paid back within six weeks from my sales for that first year.

One of my books ended up with a fellow in Hawaii. He reordered the next year and I found out he managed one of the oldest ranches in Hawaii, and if I would ever like to come there to photograph, he could set me up with about four months of shooting. At that time it was just a dream, but a year ago last November I flew over to Hawaii. He toured me around and I visited four islands.

And it was very exciting to have my work on the cover of *Western Horseman*. Then I had the opportunity to meet John Garden, who was doing a book on the Palliser Triangle. He happened upon my work in the *Canadian Cowboy* magazine in a 7-11 store in Revelstoke, B.C. He just picked it up and was browsing through it and was looking for some ranch-type photos that documented the Palliser Triangle area, so he called me, came out to visit me at High River, liked what he saw, and purchased some images from me. Within a year he had the book ready to go to print.

I see the ranch woman as being the pillar and the post of the ranch family, and I think that has been true for generations – the kind of person that binds the family together. Ranch women are probably one of the most versatile people on earth, mainly because there really aren't many jobs they can't do if they put their mind to it, whether it be sewing, pulling a calf, driving a tractor, fencing, wiping a tear from a child's eye, or cooking for a crew on really short notice. So I think versatility is probably a key qualification for being a good ranch wife.

The outlook for ranching does seem scary what with all the rules and regulations and the anti-branding, the oil and gas exploration, and now BSE. Every time ranchers turn around, it seems there's a new angle or someone trying to step in and make it more difficult for them. A lot of people I meet are really dedicated. I guess they have that true pioneer spirit I talked about earlier. No matter how tough the going gets, they'll just get tougher.

WEST OF THE ROCKIES, KAMLOOPS

Kamloops is situated in the interior of B.C. on the Trans-Canada Highway beside the South Thompson River. With an elevation of 1,100 feet (335 metres) above sea level, the region is semi-desert with coniferous forests on the mountains and excellent summer grazing pastures in the high country. There is good winter grazing in the irrigated meadows of the low country, and irrigation produces two or three hay crops each year. Winter weather lasts for only about three months, and summers are long and hot.

Kamloops was one of the earliest ranching districts in what was, in the late 1800s, the Crown Colony of British Columbia. Thadeus and Jerome Harper started their four-million-acre ranching empire just east of Kamloops along the Thompson River in the 1860s. This became their home ranch for wintering the herd because of the mild winters. Ranching remains a major industry in the B.C. interior, along with lumbering and tourism.

PAT KERR WITH THE OLD MAN

PAT KERR

Whatever you see in the movies is definitely not what a rancher's wife does.

Pat Kerr and her husband, Raymond, lived on and operated the picturesque, historic Harper Ranch just east of Kamloops, next to the Kamloops Indian Reserve. Their children and grandchildren lived and worked on the ranch with them: daughter and son-in-law Vi-Ann and Keith Nowiczin and their sons, Destry and Harper; and son and daughter-in-law Norman and Danielle and their children, Ty, Rachel, and Samantha.

Raymond died of a heart attack in 1999, shortly after making a handshake deal with Manny Jules to sell the ranch to the Kamloops Indian Band.

I WAS BORN in Vancouver but brought here as a newborn infant to the Kamloops area. Actually my home was two miles from where we sit today, on the south side on the banks of the South Thompson River. I was the oldest of four girls.

My father was a vet from WWII. After the war, he worked for the Ford Motor Company his entire working life. My mother was a homemaker. I was raised in the Catholic faith, which I don't practice much anymore, but I went to a Catholic school my entire life. It was a co-ed school, St. Anne's Academy, and it was one of the first schools in the province to do integration, where they brought the Native children from the residential schools and sent them to high school with us at the Catholic school. It was an experience. I left school in the tenth grade and took what they call a commercial course to become a secretary/bookkeeper, which basically is what I did before I was married.

I first met Raymond in 1961, during 4-H steer show time. We were married in 1964 – I was eighteen years old. All my dad said to me the day I left home was "You made your bed, you lie in it." I was the oldest of four girls who had a taskmaster for a father, and being outside is what I did best. Being in the house was not something I liked, I don't even care for it today, but I probably could build the best barbed wire fence of anybody. My dad felt that idle hands got you into trouble and us girls were never allowed to walk the streets or go to town unless it was for a specific reason – he kept us busy at home. We had horses and my mother always had a milk cow, and it was just ... responsibility.

After we were married I rode, but there were men to feed and canning to do. I lived in a little trailer down the road and my in-laws were in this big house, and we all shared the work – it was just part of it. And we didn't have all the modern conveniences that the young ranch wives have today. I mean, men had to be fed three times a day, seven days a week, and wherever you could fit in and help, you did it. It was just expected of you.

Unfortunately, Raymond became the boss overnight when his father died unexpectedly two years after we were married. It was a very trying time for all of us because he not only had a wife and a baby to take care of, he had a mother and four siblings. And then shortly thereafter his uncle, who was part of this company, died unexpectedly, too, so now all of a sudden Raymond was responsible for all these people, but he had broad shoulders and he did it well.

In those days on the ranch, the men did not do any inside work and the women did not work outside, except for barn chores and the gardening. We always had a Chinese irrigator then. When I first came here there was no such thing as pipe irrigation – it was all ditches. The garden was put in by a Chinese man named Chou En, who actually worked here as a very young man for the Harper brothers just before their demise. Chou had the garden and he planted it, looked after it, and we harvested it.

It never entered my head to have a job other than being a mom. And I always had a job here, there was never any want for work around this place. Whatever you see in the movies is definitely not what a rancher's wife does. There is no nine-to-five job, there is no telling you, "It's okay, you don't have to go to the barn this morning because you've got a sore toe," or whatever your problem is. The cattle have to eat or the horses have to be fed, the milk cow has to be fed or the chickens. They are solely dependent on you as a caregiver of these animals – yes, and the men too. They worked very hard, but they also played very hard, when it was time to play. We played together – it was not anything that we didn't want to do. We did it together.

Irwin Kerr, my father-in-law, was a very smart man. They all believed, the whole Kerr family, and my husband right up to the day he died, that there could only be one boss. It doesn't matter how many partners you have, there could only be one boss, and basically that's how this ranch has operated all these years, for thirty-seven years. Since Irwin died, Raymond was the boss, and that was it. Sometimes I would think I was, but then I'd definitely have my wings clipped. The only time I was the boss was when I had to deal with people in and out of the cookhouse because I was the one who had to clean the mess up. I could tell you some stories about ranch cooks that would curl your hair. But if you had a good one, you treated her right and she treated you right. If you had a bad one, let me tell you, they were bad.

Raymond was a great one for making sure that his mother and his aunt understood where the ranch was going and whether there was money at the end of the year when payday came. In any major thing, the women were a part of the decision-making. He always tried a couple or three times a year to have a meeting with them, and he was very good about keeping them informed. Unfortunately for his mother, Violet, she chose to leave the ranch, and in hindsight over the years we should not have allowed that. She lived three miles from here on her own little acreage where she thought she wanted to live, but she became very lonely and bitter. Here on the ranch it's like a little town and you go next door for coffee. Auntie Madge came every day. My kids loved to go to her house to get chocolate chip cookies, which she's famous for. And for a long time Violet felt that she was not part of it, and she would imagine that things were happening that weren't, and sometimes it was very difficult to deal with that. Raymond and I often wished that we had been stronger and not allowed her to move away. I think a lot of the family-related problems that we had over the years were because of that decision, and if she had lived in the orchard, as we call it, I think it would have been the other way and I don't think we'd have had all the emotional trauma that happens sometimes in families.

Raymond and I, we shared everything. I would do the paperwork, and also we worked – you know, I had to work. Ranches go in seasons just like the weather goes in seasons. Spring is new life, the babies are born. Summer is the growing period, fall is go to market, winter is be the caregiver and take them over the winter, and basically that's what a rancher's life is all about.

A typical day in the winter – if I didn't have a cook in the cookhouse, the men came to my own home. I live in a huge old ranch house and it's got a huge big kitchen. This would be when my children were in elementary school. Breakfast was on the table at six-thirty, and remember, it's not cereal and toast. I had to be gone by eight o'clock to get the kids to school. We lived on a gravel road, ten miles of gravel there and ten miles back. I probably would have been up since five because in this house for many years we only had one bathroom – my greatest dream when my children were growing up was to have my own bathroom and my own hot water tank, because if I didn't get to the shower by six, I never got any hot water. We often laughed about that. I finally got the shower and the hot water tank when the kids left home. But anyway, I would be gone by eight. I would already have a roast in the oven or whatever I was having for lunch on the go. Usually I would drop the kids off, pick up the mail, usually had to stop at the equipment dealer to pick something up, be back here no later than ten in the morning to have lunch on the table at noon. The men would come in usually at noon – they had to be on time at noon. There was no excuse to be late for lunch as far as I was concerned, unless they had gone a far way and they had a flat tire or something. Clean up, organize my thoughts for supper, leave here by 2:15, pick up school kids anywhere from 2:30 to 3:15, depending on what was happening. My daughter did take piano lessons – I made sure she was able to do those things even though we didn't live in town, and sometimes it was a struggle, but we managed it. We were very involved in 4-H, horses. Then it was supper on the table, homework, bed – TV was not allowed in our home during the week.

I guess every day on a ranch is different. Sure, you're geared by the clock for feeding the men or doing that kind of stuff, but every day is different – there is a new challenge, or a new something. You work hard, and being a rancher's wife is a way of life. It's not for everybody. Lots of marriages do not survive because of the changing, the things that go on. There are challenges every day.

The biggest thing I had to learn coming here as a young bride was that I don't want to know what's going on in your house. If you're having a fight with your husband, that's your business, and if I'm having a fight with Raymond, it's my business. I don't want to know that my husband, who is your boss, has just chewed your ass out. I couldn't care less, it's got nothing to do with me. We had anywhere from three to five families living here on the ranch, depending on the season or whatever, with and without children, mostly with children. And that was the other rule that we had. We didn't encourage the children to go into each other's homes, and it worked really well that way. I mean, sure, there were times when they all came and we all did things together, but I just found that that was the best way to handle so many families living here. I was also adamant about my kids showing respect for the people who worked here – we were all here for a reason.

I was involved with a political party for a while and I believe I was instrumental, with another friend, in developing polling by telephone – it was what we called telephone banking. It was what we started and wrote the book on. It was a fun thing to do and it's still used by parties today – they copied it from us. I still keep in touch with those people today. We've all gone on to other things in different ways. We still, I think, have a little yen to be involved in politics, but it is so electronically controlled now that they've taken a lot of the old-style politics away. Then there's the media and the fact that they are ruthless, and there's a lot of us who don't want that to be part of our lives or to interfere with our children or grandchildren. We've just backed away – it's time for a new breed to deal with it.

I also got involved with the hospital foundation, which is the organization that raises money for equipment. Through their charter, you're allowed to spend nine years there, which I did do, and I chaired a couple committees. The foremost was the finance committee, which was a real challenge for me. We went through a couple of campaigns where we raised over ten million dollars. And for me – who never finished high school, just got a bookkeeping certificate – to be the chairman of this finance committee ... well, when they asked me to do it I was absolutely terrified, and my husband said, "Never say never, go and do it, you can handle it." And it didn't take me very long to realize that you just surround yourself with the right people and they make you look good. So I had two chartered accountants, a lawyer, and a man who was very much involved with a big lumber company in B.C., and we became great friends. It was a learning experience as much as anything. If I had to change anything about my life that was within my power, I would probably finish school.

I worked outside with the men sometimes, too, depending on the time of year, depending on how much help we had. Horses and me, they're my affinity and I always have them around. I always wanted to spend more time riding and more on the cow end of it than I was ever able to do, but when you don't have a cook and kids have to go to school and you're it, that's what happens. If hay needs to be baled at ten o'clock at night, well, you just go do it. I could never do that when my kids were little because how do you leave them? But as they got older, if there was hay to bale and nobody was here to go do it, you just went and did it. If the fence is broke down, you fix it. I guess, for me, there wasn't a job here that I couldn't do – I made myself know how to do it. I usually had the night shift when the heifers were calving, but I wasn't ever perfect about it. The only job on this place that I never learned to do was run a swather. I was terrified of it. It didn't matter how big and fancy it was or what it looked like, I was not having anything to do with it and I never did.

I think the best part was the things we did as a family. Both kids were involved in high school rodeo. We rodeoed hard, we played hard, and we worked hard. We needed to get our work done in order to go to a rodeo, and if the work wasn't done, we didn't get to go. It was plain and simple, but we did it as a family – we did everything together.

This last ten years both of my kids have been very involved with the ranch. I quit doing the books, that's not my job anymore. I have a smart daughter and she took it on. My son was outside working. He helped manage with the cow boss – he learned the groundwork of running this operation and I often think he was very lucky. One day he was feeling sorry for himself not too long after his dad died, and I just said to him, "Excuse me, you had your father ten years longer than he had his, and in that ten years he gave you everything that you needed to be successful, so get off this chair and go to work. That's the bottom line, go to work and you will survive."

I always believed that we didn't have a son and a daughter, I believed that we had two children. I believed that they should be treated equally, and they were, but it goes back to what we discussed earlier that there can only be one boss. And when Raymond passed away, Norman became the boss. Period.

Vi-Ann's husband's not a rancher, he's not a cow man. It's not his thing. He supports everything we do, he's probably the best son-in-law in the world. I'm so fortunate. And it's the same with my daughter-in-law, I'm very lucky. She wasn't raised in this world – she was into theatres and arts – but she's a wonderful mother and takes care of my son, and I can't ask for more than that. It's not a bother to me that she's not a hands-on rancher's wife. Not everybody can walk in my shoes and I don't want anybody to. That's not fair.

Between ten and twelve years ago our plan for the future was – Norman had just finished college, had just got his degree in animal husbandry, he was going to come home, he was going to work for his dad. They were going to carry on raising cows. My daughter and her husband, after living off the ranch for six years, had moved back to the ranch.

But ten years ago we became embroiled in a very sad Indian land-claim situation, and that made us all rethink where we were going. It was a hard thing to live with. The first claim involved a piece of land that had been deeded in 1872 from the Crown Colony of British Columbia to a man by the name of James Todd, who in turn sold it to the Harper brothers. We tried to develop it, but ultimately they took us to the Supreme Court of British Columbia. The trial judge said it was a triable issue, and in essence the land was frozen. They basically took the right-of-Torrens system, which said we owned the land, away from us. The courts did. We almost lost everything that we had. Out of our own pocket it cost us over a million dollars, and really we got no compensation from any level of government. As soon as the courts gave that piece of land to the Natives, we received a letter from Chief Manny Jules informing us that they were claiming the rest of the ranch.

Raymond and I spent many a night pondering what to do. With the history of the ranch, it was a heart-rending decision to sell, and when we made the decision to allow the Kamloops Indian Band to purchase this holding, it was strictly a business decision – we did not allow emotions to be involved. When we looked at the overall picture, of how the ranch was encompassed on three sides by the Band, by land that they had claimed, we knew that there was no future here for us, or our son, or our grandchildren. We knew that because of the way the politicians think these days, in all political parties, at all levels of government, that nothing was going to change, and the Band was going to end up with the land anyhow. It would have been piecemealed out to them and it would have broke us trying to hold on to it, and in the end our children would have had nothing.

As we understand from the documents that we've read, the Band intends to take the title of the land as we know it and turn it all into reservation. It's a five- or six-year process with the federal government. What's happening is we have no political people today who have any guts to stand up and be counted. They make all these wonderful promises, they do and say what they have to to get elected. And we as the taxpayers keep paying. And the taxpayers of this country bought this ranch no matter what they say, that's exactly what happened. Our home is gone now to the Kamloops Indian Band. I figured I was going to spend the rest of my life here. That's not going to happen now. It's sad to see a ranch, probably one of the most beautiful places in Western Canada, be gone from what is known as our western heritage. The Harper Ranch was barely economically viable with the cost/price squeeze in the ranching business today. It was family tradition that kept it running.

You know, I could be bitter, I could be nasty, I could be ugly, but what's that going to accomplish? Nothing. It was a decision that we had made together, so I'm fine with it. The fact that I have to leave my house is not fine, but a home is where you make it, so let's just get on with it. It's hard for my children – they've never lived anywhere else – but as I tell them, one door closes and another one opens and it will all come together when it's time.

I don't see much of a future for ranching in this province because I feel that we are totally controlled by the Agricultural Land Commission. Many ranches in this whole South Thompson valley are gone today because the banks would not help them. They took ranches and turned them into hay farms, and with hay farms it's feast or famine. The B.C. government as we know it is not agriculture-friendly at all. It is very environmentally friendly. They have forgotten that somewhere along the line we need to feed the people, and they are not looking at that. They don't want any more logging, so consequently where's the money going to come from to pay for the health care in this province? We hear about the health care crisis every day. Agriculture is probably in the lower five percent of concern by this government that we have today.

If Norman decides to buy a place to go back into the cow-calf business, I would want to be involved. I'm not ready to retire. I know how to clean a barn, cook a meal, fix a fence, drive a tractor, buy fuel, calve a cow. That's what my education is. Maybe I'll keep the horse end of the outfit and he'll raise his cows. He is the boss, and I want it that way. I just want to be around. And I have other interests, too. I help raise money for the crisis fund in pro rodeo, and I personally am involved with the High School Rodeo Foundation, which is the arm of high school rodeo that generates all the scholarships for high school students. We have auction sales and we do all kinds of things. I also am on the Miss Rodeo America committee.

But the land sale was not the only thing we had to deal with as a family. Skeletons in closets are part of every family, and we were not immune from them. Child abuse was one issue we dealt with, causing a huge family rift, never to be healed. Over the years the pain has faded, but it has never gone away. And then, over the past ten years I have nursed four loved ones through cancer to their deaths – my father, my mother, my mother-in-law, and my friend Lois. My father-in-law and my uncle-in-law died suddenly many years ago. Heart, just like Raymond.

Then there was Frankie. Well, let me say that Frankie was probably the best brother-in-law that anybody could have. He was the next oldest in the family to Raymond. He was

Raymond's partner and his best friend, and Frankie loved life, he loved everybody. We always believed the police would come to the ranch door and tell us that Frankie was gone because he'd wrapped his truck around the nearest telephone pole, because that's what he used to do. He had nine lives, that man, and for him to die the way he did was just totally unacceptable. It was so out of character, but I guess it was his time, they had room for him that morning. He died in a fire, tragically, on New Year's morning. He was twenty-five years old, had the world by the tail, could model for any magazine, was so good-looking. Women stacked like cordwood outside his door, but a best friend. So that was really a hard thing for Raymond to get over, you know – Frankie and him had shared so much, and then he had to walk alone. It was tough.

I honestly don't know where I got the strength to get through some of these things. My father was a taskmaster, and I guess I just had to stand on my own two feet. I didn't have anybody else to stand on but me, and I've just always done that. I think that you gather strength from your partner, because Raymond was strong, too. I guess that's part of the strength – we just shared it and did what we had to do. I guess for me, and I think I can speak for Raymond even though he's not here anymore, we never had the luxury of locking the door and throwing away the key. When someone died there were still cows to be fed, there were people who relied on us to make sure they got their paycheque at the end of the month, cows had to be milked, chickens, horses had to be fed. Some people have a death or an accident and they feel sorry for themselves, but that luxury was never there for us. We had to keep going. Lots of people are worse off than me. I'm very fortunate that my children are healthy and my grandchildren are healthy, so I can't ask for anything more than that.

I don't practice my Catholic faith, but I do believe there is a God, and I don't know who the hell She or He is but there's somebody who definitely helps me lots of days. And I also truly believe I'm not the first person to walk this road and I'm not going to be the last. Just accept it and get with the program.

When Raymond died I didn't plan anything, but I know we had to have something, so I said to the kids, "I don't want it called a funeral because he hated them. You can call it a gathering. Just say friends and family welcome at the ranch at one o'clock on whatever day you choose to have it." We had a family friend who had been around a long time. He had been an inspector in the RCMP, and he gave the eulogy, and I said, "You have exactly five minutes, that's all you have." Raymond's auctioneer friend, Donny Raffin, kept the gathering a time of celebration, not sadness. And I said, "Norman, get the beer truck here because that's what he'd want." So, that's what we did. As it turned out, they drank eight kegs of beer that day. There were over a thousand people here on this lawn. Raymond had many friends from all walks of life, and they shared all their stories about him that day. Even through all the sadness, my children were very proud of their father.

I've had so many losses, there's no clear picture for me anywhere at this point. I think it's one day at a time, and there's nothing I would like better than to stay in the cattle business. I don't know anything else. At this point I'm just going to move to a property not far away, that's the plan, see what spring brings, and then go from there. It just takes you a while to sort of find a new road to walk because you have to walk it alone. But I've dealt with death so often that I'm not afraid, and that's probably what's made me survive.

October, 2003

Pat Kerr lived for two years in a house rented from the Kamloops Indian Band, along the South Thompson River, on land that was once the

Harper Ranch. She has recently nursed another dear friend through cancer to her death. Presently, Pat lives in a condo in Kamloops, and intends to build a new home in the future on some land she owns along the river. It has a view of the valley where she has lived all her life. Her children have bought property thirty minutes away in the Cherry Creek area, and the family is gradually returning to the cattle business. It is also interesting to note that Norman Kerr has recently been hired by the Kamloops Indian Band to be the manager of what was formerly the Harper Ranch.

SOUTHWEST ALBERTA FOOTHILLS

The southwest corner of Alberta, from Stavely down to Waterton Glacier International Peace Park, encompasses rolling grasslands and forested foothills bordered by the Rocky Mountains. Three major rivers wind through the area: the Old Man, the Castle, and the Crowsnest. Chinook winds provide relief during the winter months, with temperatures increasing as much as fifteen degrees Celsius in one day.

The North-West Mounted Police travelled to this area in 1874. They brought 235 head of cattle up from Montana that same year. The lush grasslands along Pincher Creek also made it an ideal location for a horse ranch for the NWMP. Methodist missionary John McDougall and his brother David brought the first herd of breeding cattle to the area in 1873. Many of the NWMP began ranching after they finished their terms as policemen.

By 1882 nine thousand head of cattle were established in the area. Large ranches such as the North-West Cattle Company – later known as the Bar U – west of Nanton, the Walrond Ranch north of Cowley, and the Oxley Ranch west of Claresholm were operating. Stavely, Nanton, and Cowley became the service centres for these ranches. Auction markets sprang up where ranchers could sell their cattle to be shipped by train to the East or into the United States.

Along with the men, the women came prepared to face hardships, isolation, and in many cases, poverty – in order to begin a new life in the West. Small homes often miles apart were tucked in the valleys or trees where ranch families eked out a basic living. A trusting relationship developed among townspeople, ranchers, farmers, and their families.

HELEN CYR

I'd be out there more if I could or
if I had somebody to do my work inside.

A mixed farm near Longview, Alberta was Helen Cyr's childhood home. The second oldest child of seven children, Helen was born to Jean and Gerald Herriman on February 8, 1951. Today, Helen and Clarence Cyr live on their Valley Blue Ranch, Half Diamond XU, ten miles southwest of Pincher Creek in the foothills, where they raise Parthenais cattle. They celebrated their thirty-third wedding anniversary in September, 2004.

I LEARNED HOW TO milk cows and I wished I hadn't ... and I drove grain truck, too. I was the tomboy – I liked the outside work, riding and working with my dad until my brothers got big enough to help. And we had a farm, so there was pigs, chickens, and milk cows, just a variety of chores. We were taught to work hard and enjoy it or else. They wanted us all to be hard workers, pay our own way and make our own livelihoods. That was important. We really never talked about staying on the farm or staying in agriculture. That aspect of it never came up. Dad actually left the farm to my brother.

HELEN CYR WITH FAMILY, TAKING A BREAK AT THE CABIN

We never missed school because of harvest. Mom and Dad didn't believe in that. I would drive truck after school, hauling grain, and do my homework in the truck. As far as further education, Mom and Dad didn't have money to send us off to secondary school, so we didn't unless we did it on our own. I did go to Olds but that was to upgrade my high school marks, and that's where I met my husband. Since then I have taken a university course. I was a unit clerk in a hospital in Edmonton and a teacher's aide for a while when my two oldest boys were really little. I enjoyed that. I am sorry now that I didn't get more education. I wish that I had something to fall back on, but a lot of things that I have learned ranching – doing books – and dealing with people – have been good training for any kind of a job.

The first year we were married after I had worked in a busy hospital, we were snowed in half the winter. It was pretty hard to take the loneliness, and not knowing a lot of people down here except family.

My father was a little pushy. He wanted grandchildren but I didn't let it bother me. I think I was twenty-one. The first two sons were born in the Pincher Creek Hospital and the last two were in Lethbridge. We had gone down to an Alabama concert the night Jeff was born, came home, my water broke and we had to go back to Lethbridge. It took over an hour to get back to Lethbridge but we made it. With the first two, Clarence figured it was just like a cow calving, so he didn't want to be there. For the third son the nurse pushed him into a gown and said, "Here, you're going in there, aren't you" and he was quite thrilled. My mom came down and stayed with me every time, so that was a big help.

Irene Reeds and I have been friends for years. We had kids about the same time and we were tied down at the same time, so we'd phone each other, or they'd go for a Sunday drive and stop by and see us, or the other way around. That was really supportive. I found it hard. I was a doting mother. Sometimes it was harder because I found it hard to get a babysitter for them. I would drop them off at the neighbours if I had to go to town, and she'd do the same, but not very often; I usually took them with me.

The first three were two years apart and then there was a gap of seven years. I figured if you got the boys the girls will come. I did lose a baby in between the last two boys. I rode more probably with Paul, too, than I did when I was pregnant with the others. I didn't want my freedom to become too limited. I can remember driving tractor with a toddler and riding with a baby, moving cattle or just pleasure riding, taking the baby along. I loaded one on a horse and one would sit behind me and I'd lead the horse and off we'd go. It was fun.

My role was in the house, cooking and cleaning. I don't enjoy dusting a whole lot and cooking's not my thing. I do it because I enjoy eating it. Clarence didn't share my work that much though, and he still doesn't except if I get on his case about it. I don't think he knows how to vacuum. He can wash dishes, especially when he wants to get his hands clean or when I really fuss and say it's about time and you haven't done it for the last year. But I know that he's busy outside too, so there's got to be a division to a point.

I'm the one who looked after the kids, put them to bed at night, 'cause their dad was still out working. When he was around he shared in the parenting. He'd take them with him in the tractor once in a while, just as a joyride. They helped with the chores, helped gather eggs, and we fed pigs here for a while too. We've got cows, so every morning before school we had to wash the separator. I was outside helping to milk. They help now with branding and vaccinating and preg testing, any time we have to work cattle and need more hands than the two of us.

A turning point happened in my life in September of 1989, harvest time. Our three eldest boys were at school. Paul, three years old, was outside helping Mom and Dad fix the bale wagon, a major repair job. With the main drive shaft, being disconnected, my job was to hold it up in place. I inadvertently pushed a manual control to lower the very heavy second table that lifts a whole tier of bales up to stack. The table came down quietly and quickly before I could jump out of the way. My head was caught for an instant before I pulled free. There was no pain yet, just thoughts of getting to the hospital quickly and trusting modern medicine and technology. Clarence was at his wit's end. Luckily we had a friend who was calmer. Covering the gash in my head with a towel, we were quickly on our way to Pincher Creek. I ended up in Lethbridge, where the doctors reconstructed a crushed temple sinus and stapled up the tears in my scalp. Recovery took a while, a lot of quiet time, and the dizziness finally went away. Vanity flows out the window when my life was at stake and hopes for a normal life are a priority. I was so lucky. Ever since that day my outlook on life is a little different. I try to live for each day and not take things for granted. Accidents happen so quickly. Agriculture is a hard way to make a living and probably holds more dangers than any other occupation, but the benefits of the lifestyle outweigh those risks.

I like getting outside but I don't like getting cold. We were always actively involved with the cattle and the farming until our son got big enough to take over. It's so mechanized now with the big equipment, there's only one person needed to feed. I used to go out with Clarence and feed bales and drive the truck while he was throwing them off. Right now I am feeding a calf on the bucket morning and night. My horses are also my chores. Even in the storm when I had to put horses in the barn and keep them in there, it's a chore that I had to do, to look after their health and their well-being all the time. I'd be out there more if I could be, or if I had somebody to do my work inside.

I think ranch life is easy enough. There's not as varied chores as there used to be. I don't know if that's a good change? I think that develops responsibility. I think it is a good way for children to grow up to learn to be independent and self-reliant. There's always something for them to do. They learn how to care for animals and each other because they depend upon each other. We're self-sufficient and we help each other to get through an emergency situation. I know the older boys are always saying, "You don't treat Paul the way you treated us." We've gotten tired of having all those chores too. It's not just the child's responsibility, but we have to make that child do that and it's a push.

They do enjoy those roots. They want to come home. They don't want to see it go. But as far as making a living from the ranch, somebody once said that it was child abuse to leave a ranch to your child. Tim, our son who is a graphic designer, is very interested with the way we're marketing our beef right now. Jeff, our third son, enjoys coming out and fencing. Paul, the youngest son is a good little worker too, especially if he's getting paid to do it. A lot of times it's give and take. I see a change now with Cody, our son who is working with Clarence. They talk together and make a lot of the decisions. I still expect to be part of that decision-making process, especially when it comes to spending money. Sometimes the goals are the same and sometimes they're not because parents focus more on work. We've slowed down, my role specifically. I walk instead of running or I take the quad. Our son will hopefully take it over and we'll fill in when that extra help is needed. Hopefully we can continue to live here as long as possible without having to move into town.

Our days are our own. I think the freedom of having our own place and being able to go for a hike when I want to on our own place is important. I go canoeing, hiking, and cycling too. We have a women's weekend up at the cabin on the long weekend in September. Once a year my sisters and girlfriends get together. I find that we're becoming more active, maybe it's in desperation because we are losing it. I haven't done as much riding as I would like to do. It's because of sore knees, a sore back, and other commitments, like meals-on-wheels to the field. It takes me longer to get through things than it used to.

I volunteer for a couple of different organizations. I feel that's important to be part of the community. I'm involved with the Agricultural Society with cowboy poetry, the fair and rodeo with the trophies and at the AGM. It's a tough commitment being able to go into town and do that too because we're about ten miles from town. For five years, I was on the committee to establish the women's shelter in town. I'm glad that I did that. Farm women are the more stable core in an organization. They're very responsible to do a job and do it well.

Financial worries are greater now than ever. There are so many high input costs compared to the price that we're getting for our cattle. It's getting increasingly harder to make a living from straight ranching. I see a role change because a lot of women are working outside the home too, as well as having their own farm or ranch duties. They're spreading themselves pretty thin but I see it as a necessity. I think a lot of the women do the books on the ranch. That's not a small job anymore – it's a huge job. As the ranches get bigger they will need more of their time too. I don't know how it is going to work out financially.

ALICE STREETER, JAC AND FRIEND

ALICE STREETER

There was a lot of trust and friendship.

Alice Streeter was born in the old hospital in Nanton in 1924. She had three brothers and one sister. She is the only survivor of her immediate family. Her folks, Jim and Gertrude Smith, were farmers who lived west of Nanton. The setting of the L4L Ranch where Alice and her husband Jac, made their home is a picturesque valley tucked in the foothills southwest of Stavely, Alberta between Highway 2 and Highway 22, where the Willow Creek winds through the valley. The ranch is twenty-three miles west of Stavely and thirty miles from Nanton. There are two homes in the yard – the little house, which is one hundred years old, where Alice and Jac Streeter lived when they were first married in 1945, and the larger ranch house, built in 1919, where Granny Streeter lived. This larger house later became Alice's home. At the beginning of August, 2004, Alice left her beloved ranch and moved into Claresholm.

BOTH MY PARENTS worked very hard. We just got along and helped Mother with the garden, the turkeys and chickens. We all went to a country school two miles north from our farm.

We either walked or rode a saddle horse. That school only went to grade nine. Our parents didn't have the money to send us to Nanton and there were no buses in those days. I didn't want to go to town and I'm still that way. We were all happy to be at the country school fairly close to where we lived. Jessie, my sister, and I didn't really look ahead to see what we were going to do. We just thought we'd help Mother and Dad out. The biggest interest was to get out. My two older brothers were working out and my sister, she worked out.

When I was eighteen years old, I went into Nanton and got a job at Armstrong's clothing store. I worked there two, two and a half years. That was where Jac and I met. He was coming in quite frequently to get a pair of overalls or socks or something. We were married, just a small wedding, in July of 1945 at the Central United Church at the time of the Calgary Stampede. Jac had to rope that afternoon. He caught the calf and the damn thing got up. He didn't have his mind, I guess. Anyway, we didn't get anywhere that night because we had umpteen cowboys come up to our room. Then the next day Jac and I left to go on a great big long honeymoon. We went to Banff and Lake Louise and then we had to come home and take care of the ranch. The first time he ever brought me out here, honest to God, I thought we were heading right back into the mountains. It was a long trip and I didn't know where we were going, had no idea where Streeters lived. Of course, the vehicles weren't that quick in those days either. There wasn't much in the way of roads for the last fifteen, twenty miles – just a trail from the Flying E, which used to be the Burn's huge ranch where Bill Cross is now.

Jac's brother, Allie, and his mother, Mary Ann Streeter, were here on the ranch, the L4L, with Jac and I. Jac got the brand the TL, but we always called it the L4L. Jac's dad had run racehorses and he couldn't run horse races and ranch at the same time, so the ranch finally went to Jac and Allie. We practically ran the ranch after Allie and Kate were married. Their place was just west of us.

I rode a lot with Jac and helped him with the cattle, moving them and gathering them up for branding. Then we did get help. The hired men treated me just like I was another man when we were working outside. In the house I did the cooking. Some of the men did help in the house by washing dishes after meals and drying the dishes for me. Jac did a lot too. I always had a big garden. The first few years before freezers and fridges came out, I canned. Jac and I built a root house that was wonderful for potatoes, carrots, and all our other vegetables.

Outside the house Jac made decisions and inside the house, that was mine. Just like Jac used to say, and he said that to the hired man and anyone else, "If you're not happy with the work you're doing, get out." We were always happy. There'd be days when we'd come in dog-tired at night, but we enjoyed what we were doing and the men did too. Work in the bad weather depended on what there was to be done. If we had to move cattle one day and it was cold, it was done. We couldn't leave it until tomorrow because we didn't know if tomorrow was going to be any better than today. There were a lot of cows to calve. We always tried to keep the heifers close to the buildings so there was more help for them if there was problems. We calved usually the start of April.

We had two girls, Donna and Mary, who were born in Claresholm. I was in bed for eight days but I was allowed to hang my feet over the end of the bed on the eighth day. I felt so strong, I'd feel like I could run a mile. I'd sit on the end of the bed and put my feet on a stool and they'd be full of pins and needles. I got out on the tenth day. I had a girl in the summer to help me, and my mother came when she could. Then of course Granny Streeter was here in this house too, so she could help quite a lot. I breast-fed Donna until she was eight or nine months old. I don't think it was that long with Mary. After they got on the bottle, I'd boil milk from the cow.

Donna, the oldest girl, got red measles. She had a terrible fever when she was about five years old. Jack and Bert Cochlan were heading west with some cattle and they'd stayed overnight here. Jack Cochlan and my Jac went down to the Flying E to phone the doctor, Dr. Lloyd, an army doctor from Wales, who was new in Claresholm. He got Mr. Fairburn to show him the way out here. I was petrified. Mary was pretty small and then she got the measles, but nothing like Donna. Little things like that made me worry quite a lot. It wasn't too many years after that we got a phone on a party line.

When Donna started school we had a building out here at the back, and a high school girl who had her grade 12 came out and stayed and taught correspondence to her and Kate's boy, Donny Williams. Donna did two years with correspondence. The school was fixed up nice inside. We bought some desks from some of the country schools that were closed. When Cross bought the Flying E, the bus came as far as the Flying E, but I had to take them either in the morning or pick them up at night. Then they went to Stavely school.

They helped out on the ranch when they were younger. Donna went to Calgary and Mary stayed home and ranched with us for a long time until she went away and took her nursing aide schooling. Now Donna is back on the ranch with her family.

We always could count on our neighbours and we reached out to help them when we could. I can remember back so well when Jeanie Akins, Virginia Delinte's mom, was having these asthma attacks. Jac and I would look after the kids when they needed help. Alice and Eldred Seeley lost their little boy when he was riding on a Shetland where this stallion was, and the damn thing kicked back and killed him almost instantly. I came down and got clothes and funeral things for him. We did anything we could, including cooking food, to help them out in any way. I've helped my neighbour Margaret Boulton, who is eighty-two and crippled with arthritis, with her work and branding; I cook her a few meals. That's all I've done away from the ranch is help out my neighbours.

John and Eleanor Cross and Bill Cross from the Flying E and Neil and Nellie Riley were good neighbours. We never worried when our cattle got into their pasture or theirs got into our field. There was a lot of trust and friendship. There were others who we could count on when we needed help. If we had a party for somebody's birthday, they'd all either be here or at the neighbour's. We had a lot of fun. I would chord the piano for our singsongs. Virginia Delinte and I were miles apart, but our friendship has stayed, and that's the way with a lot of them.

Jac and I travelled quite a lot over the years, after we got money to travel. We would go for a month or so down to Illinois, Colorado, and Wyoming. It wasn't always just visiting friends, it was getting away and doing our thing. We went to the King Ranch in Texas and Jac thoroughly enjoyed that.

It has been ten years, November '93 when I lost Jac. When Jac and I were on the ranch I was helping him pretty well all the time, but now I don't get the opportunity. I don't ride now but the only reason is I don't have a horse. I gave my saddle to Tanis, my grandson Jason's wife. I'm happy here. This is my home with my family pictures. I enjoy looking after my house, my garden, my flowers, and my pets.

Ranch life has changed considerably. The kids today, as soon as they are sixteen, have to have a car and they're not helping when they should be helping – mind you, that all depends on the parents too. Donna's boys, bless their hearts, have all helped Jac on the ranch. I would certainly rather see ranch life than town life to raise kids. They can't get into the trouble the way kids do in town.

For a lot of years we didn't get a heck of a lot out of a critter. Look at the way our summers have been, where nobody had much of their own hay to put up and the price of hay now. With this mad cow thing, I really feel sorry for the feedlot people and truckers. A lot of those people have got a lot of cattle in the feedlot and what are they going to do? The cattle are getting heavier and they're going to be dropped on their weight if the line ever does open.

I don't think it is a good change from the way ranching used to be to what it is today. My dear man, if he'd thought this ranch would be split up he'd be ... because he dearly wanted one of his grandchildren to take over the ranch. A lot of that stuff I just never talk about. I'm afraid the ranches will be sold off in smaller packages to get more money out of it. That's my vision of it. They want to make it just a playboy country, all of it.

DORIS BURTON

This chapter contains excerpts of Doris Burton's autobiography, her second book, written in 1995, *Babe's Sunshine: I Made My Own*. Her story is one of outdoor adventures, horsemanship and family. Although Doris passed away in the spring of 1999 at the age of eighty-seven, her daughter June Swann, who now lives in Arizona, gave permission for Doris's story to be included in this book. Doris was a true cowgirl and ranch woman. She helped build up the Box X Ranch, rode green broncs, roped as well as any cowboy, and worked and moved cattle and horses. Doris never lost her love of the seasons and the changes that happened on the ranch, – new calves in the spring, birds returning to nest in the spring, baby deer and other wildlife.

Doris "Babe" A. Burton was born on September 19, 1910. She led, as she put it, "an almost idyllic life" as a child growing up with her sister Kay and her parents, Fredrick Herbert Riggall and Dorothea Williams Riggall. Raised three miles north of the Waterton Lakes, Alberta, the girls roamed the hills and Rocky Mountains with their parents, who had a hunting and fishing guiding business. Her playmates were her sister, her horses, cats, and wild creatures that

DORIS BURTON ON HER FAVOURITE HORSE, PLUM

wandered into their camps. Babe could still remember the names and characters of the horses in her father's string. As Babe says, "It was a very beautiful way to grow up."

The following excerpts describe adventures that Babe had while travelling with her parents on their guiding trips into the Rocky Mountains.

THE NEXT DAY WE CROSSED the Kootenay Pass. Dad and Ken went off to hunt goats and I was to tend camp and the horses [Babe was twelve years old at the time]. On the fourth day in the swamp camp, I had the two seasoned horses, Bobby and Steamboat, tied up and the other three picketed. In a strange unison, the other three pulled hard on their pickets and joining the loose one, went running down one incoming trail. I grabbed Bobby and galloped off after the truants. I got ahead of them and would have turned them back, from a knoll that I was on, but as I galloped down it, poor Bobby sank into a soupy bog. I went sailing over his head and I could hear him coming plop, plop towards me. I was sure wet, dirty, and unnerved and Bobby was heaving with fright.

I would trail the horses and hope to catch up to them at Cameron Lake or surely a place called Little Prairie, fifteen miles away. [Doris rode Steamboat, the horse that she felt she was just "stealing a ride on." After a long ride, Doris found the four horses and arrived at Bo Holroyd's, Anne Stevick's grandfather, who was the park warden. She stayed the night with them and he tied the horses nose to tail for Doris's trip back to her father's camp.]

When my parents' pack train outfit left our home on June 30, 1926, with a full quota of saddle and pack horse plus their equipment for a three week trip into the Gap in the Livingstone Range, I was well trained in horse care management, rounding up, corralling, and catching. Horses tolerated me and I could go in among them as they bit and kicked each other. I could halter and bridle them safely.

That year of 1926 drastically changed my carefree life with Dad, Mother, and Kay. ... When in 1926 our outfit set up camp in the so-called "Gathering Pasture" two miles upstream from the Gap Ranger Station, twenty-seven-year-old Eddie Burton arrived as the last tent peg was pounded in.

Eddie kept after my parents and they very reluctantly gave their consent. [Babe was only sixteen at the time.] Eddie and I were married, but I had to return with Dad and Mother as Eddie had no home or job yet. The North Fork Stock Association hired Eddie again and he got permission to use their cabin on a year-round basis. He loaded up me and my trousseau, which Kay did most of the sewing for, and my black cat. We took off for the seventy-five-mile trip to our home in the mountains. Mother and Kay came in Dad's new car loaded with good things for use in our new life.

The cabin had one layer of boards for a floor and had lots of holes through which the ground could be seen. The cabin had a peaked roof that gobbled up the stove's heat before the floor area received any warmth. The table and the two-decker cupboard were made up of King Beach Jam's wooden cases. The washstand was a big wooden case that used to be full of Royal Crown soap. There was an old-model stove with two side-opening oven doors and a hearth with a slide draft. The cabin was so cold that the bread dough my mother taught me how to make would not rise.

Dad and Mother gave me a horse named Pansy and a saddle, a camp bed and a pair of grey four-point Hudson Bay blankets, my trousseau pillows, of which I was so glad because the pillows that Eddie got from the ranch were old and very greasy. I was terrified of bed bugs in them or in the old mattress and blankets from the ranch. There was no place to display Kay's artistic crochet work.

I did truly love Eddie, in spite of how my adult life began, at just past sixteen years. He was really a clever person. He could invent things we needed out of seemingly nothing and worked hard at earning the dollars that we needed.

Our first summer went by and in September, I developed stomach pains. My mother insisted that I must get out of these mountains to see a doctor. We arrived at the Old Man River crossing, but there had been a cloudburst somewhere and the river was too wild to dare to cross. We returned home to the cabin and slowly I got better. Later, I was ordered to stay in town. Finally I had a very complicated delivery of a seven-pound son, Robert Edward, born October 21, 1927. I had never seen a baby changed or nursed, so I had to be taught by Sister Patricia. Mother slept in a tent at their camping place behind our cabin. She was a great help to me for I was afraid of handling our little son. Both he and I had developed heavy colds.

Eddie's summer wages were gone, spent on groceries, car repairs, the doctor, and the hospital stay, so winter wages had to be found for us. Eddie liked trapping and his furs were clean and well stretched and buyers preferred them at the fur auction.

I lived through a painful winter as all strength left my back, and to lift Bobby and do all the chores for him and ourselves was agony. Eddie said he couldn't get me out or get a doctor in, and he didn't have any money either. I was a skeleton with clothing hanging on me. On Dad's second guiding trip, he had a famous New York doctor, who came with Mother to examine Bobby and me at the cabin. He told me that I was depleted of vitamins and minerals. I was still growing but the child had been given the nutrition. The doctor wrote out a prescription that did wonders for me. Eddie hated the cost but it was a lifesaver for me. Bobby got better too with a regimen of supplementary feeding.

A year and a half later I was helping at all outside work. I could lift fifty-pound salt blocks onto a bronc. I salted, rode fence lines, or moved cattle away from the drift fences to better grass. I had to ride saddle horses that Eddie said needed gentling or bridle wising. How I ever got through those years without a horse breaking a leg or hurting me somehow – I always say God rode with me. With a free cabin and two venison bucks each, we got through those almost moneyless years. Mother and Kay borrowed Bobby often, so I could take a man's place helping Eddie.

I mended enough by five years to have our second child, our daughter, Dorothy June Burton, born July 16, 1933. Again I was very ill and put to work too soon, but this time I was older and wiser and had that doctor's warning to go by.

In 1933 Eddie had to have his appendix out and Kay came and took care of Junie so that Babe could do the cow gathering.

We got that big herd delivered to Maycroft, missing only three head of Waldron Ranch yearlings, too wild to round up until Christmas time in deep snow. I had to gather our horses to go out to a place called the Gallway Place, twelve miles from our cabin. I found the bunch and in a real blizzard, started them for home.

We lived eleven years in the Gap on the North Fork Stockman's Association herder's job and Eddie's trapping in winter. Our first five years were what I look back on as family years, just devoted to the business of eking out our living with the meagre human comforts that we could afford, little by little, such as battleship linoleum on our twenty-by-sixteen-foot slivery cabin floor. We papered our steep peaked roof inside with the wrapping paper and flour-and-water paste.

After riding the rough areas in the mountains and the Gap for eleven years, I felt like I was riding in a park when we moved out of the Gap to the Sitton Ranch, twenty-three miles west of Claresholm on the east slope of the Porcupines, at the bottom of the Seven Mile Hill. By April 1938, Ed took over the running of that Sitton place. But just as the kids and I were about to move, the Sitton house burned down. We moved into a garage for three years. In April 1941, Ed put a carpenter from Hungary to start on a house on the foundation of the former burnt house. Then a big new barn was also built, a blacksmith shop, two more garages, a shed for light plant batteries, a new hen house, and a tiny bunkhouse.

Work on the ranch was far more demanding, though I loved it all. Ed was a slave-driver and hard to please, but I understood the stress and tension and did my very best. We raised palominos with a stallion called North Star. We had Hereford cows and five black Angus bulls. Bob helped until he was seventeen, then he left. Junie loved horses and became a good roper and horse breaker. We bought the Sitton Ranch in 1940 and renamed it the Box X. We heralded in a happy, peaceful, middle area of our lives, full of love that was sincerely caring one to another. The ranch duties were tackled without the taskmaster, overlord atmosphere.

Ed had hunted for years and I had hunted about ten years. We found cougar-hunting paled all other hunting. [There was a $40.00 government bounty on cougars]. We graduated from shooting to capturing for the Calgary Zoo and Al Oeming, who took two or more of our captured cougars.

It is sad to say, but that very peaceful 1945 to 1960 span of years, finally ended unhappily, but I am so very grateful for the fifteen years of the kind of marriage that I had always hoped for. I first noticed when these happy years were ending because Ed started beating any animal that I liked.

One year when I returned to the ranch after surgery in Rochester. I found the ranch, in Ed's possession at ten minutes past midnight on my fiftieth birthday. I went to a lawyer – when I got a registered letter from Ed denouncing me. After Ed died, I went back to the Box X, looking after it for our two granddaughters, who inherited it. They were underage and living in the United States, so I took on its overseeing until 1975 when the youngest was twenty-one and I continued until 1981.

Doris then trained as a nursing aide and lived in Claresholm and later Pincher Creek. Until the end of her life, Doris continued to enjoy her journeys into the hills, having tea with friends, story-telling, and interacting with people and animals. She last rode a horse when she was eighty and she branded her own cow when she was eighty-two. She wrote in her diaries each day, working on organizing her photo and film collection from her father's hunting years and her own life. She wrote and published two books and was working on a third when she passed away. Doris was predeceased by her son, Bob.

VIRGINIA DELINTE

I think women probably worked harder than men did.

Virginia Delinte was the daughter of Jean Stewart and Ted Weiser, who died when his Navy ship, the *Margaree*, sank in the Atlantic in 1940 during WWII. Her mother married James Edward Akins and when her mother passed away, her stepfather married Edith Hewitt. Virginia was born on December 8, 1938, in High River, and her family moved several times from near Saddle Mountain north and west of Chimney Rock, finally settling north of Cowley, Alberta.

MY MOTHER WAS sick a lot with asthma, so every time she came in contact with certain animals she would have an asthma attack. In those days the only way they could help her was to give her a shot of adrenalin to get her heart pumping. When we lived out west of Willow Creek, Dad would put her in the car and we would take off to town. We didn't have very good roads. They would drop Barbara, my brother, and I off at one of the ranches like Alice and Jac Streeter's, and then Dad

VIRGINIA DELINTE

would take Mom in to the doctor in Stavely, Claresholm, or High River to get her some oxygen. In 1947 after we moved to Olin Creek, my Mom had another baby, Bill. My Mom died in February when she was twenty seven and I was nine. She had an asthma attack where we lived out north of Cowley. Susie Hewitt, Ed Hewitt's wife, took Bill and looked after him. We would ride to school and then stop in and play with Bill so he still remembered us. We had a succession of hired girls looking after us, and then Dad ended up getting married to a young woman named Edith Hewitt. We had more brothers and sisters after that.

I remember there for a time when Dad would say the place for women is in the house and the place for the men is outside. But then as he got a little older he got so he couldn't do a lot of the things. It was up to us to help hay, thresh, chase cattle, and brand. I think the most horrendous thing I ever remember as a kid was helping with dehorning. We had Herefords and we had to dehorn. There was blood everywhere. I felt so sorry for the cattle. I can still remember crying. It had to be done before the green grass came, otherwise they would really bleed, so we used the iron and pine tar to seal it to keep the flies away. As technology got better we would cut off their horns, then use the caustic soda.

We rode a couple of miles across country to Olin Creek School, on the north end of the Snake Trail. When they closed the school, my Dad didn't want us to be boarded out, so we moved up to Spondon, which was northeast of Hanna. It was a one-room school and I'd listen to the teacher teach everybody the subjects. Some of them I concentrated on and got my own work done, and then I listened to the rest of the conversations.

My Dad was a calf-roper, so rodeos were our social life and our holiday. We went to Taber, Medicine Hat, Lethbridge, High River, and Calgary. There were always one or two rodeo fellows that would be down and out and they would come and work for Dad for room and board. There was one old fellow named Pedro, who taught my brother and I how to chew tobacco and spit, and drink out of a jug over our shoulder.

I graduated and got my Nursing Aide Diploma. Hubert wanted to get married but I said I wanted to work. "No wife of mine is going to work." "I thought, "Oh boy, here we go."

When we first got married, we worked for the Armstrongs out near Eastend, Saskatchewan. Hubert rented out the land and cattle. We just thought it was a good idea to get away from everybody and get to know each other. He was gung ho to go but he got so homesick for the mountains. We got married in April and came back in September. I think women probably worked harder than the men did. I was very strong-willed and sometimes we really locked horns, but Hubert always used to say that I made it so much harder on myself than was necessary.

I had six children, Deb, Carolyn, Brian, Sharon, Keith, and Greg, all within ten years. I had a lot of problems with the contraceptives they had in those days. I'd gone down to a doctor in Lethbridge and had to go before a board of doctors before they would sterilize me. There was four or five of them and they sat there and quizzed me about why I wanted to do this. Then I had to have my husband sign for me. I didn't have control over my own body and that made me furious. He thought it was a good idea, but the doctor and the board kept asking him why. We had five children already and he said, "I think that is enough." They agreed with him. Then they put me off for a month, then another month. In the meantime I got pregnant. I went down to have the operation but when I woke up the doctor said I didn't have the operation because I was already pregnant. I was mad at the doctors and mad at my husband. After I had the baby, I was fine. The doctor said,

"As soon as you have this baby I will tie your tubes." It was stormy, a terrible January here, so I went down to Lethbridge and stayed with some people we knew, and had Greg there in 1969. They still needed permission to sterilize me, so they had to do it over the phone because Hubert was snowed in.

The childhood life of the kids went so fast and I was so busy. While the kids were little, we had to pack, heat, and throw the water out. I had to see that everything was fed, children and animals. I like cooking and baking. We had pigs, chickens, little bum lambs, and milk cows. During calving I had extra laundry like you wouldn't believe. In the wintertime it would snow and snow, and I'd have cabin fever way out there and the roads were awful. We lived thirty-four miles out, twenty-four from Cowley. I tried working different times but I just got over-tired and ended up sick 'cause I was too wore out and just couldn't function. When my youngest boy got to be four or five, I got my license and drove the school bus so that I was home with them. In between we would do things that had to be done on the place.

I think the ranch is the best place for kids to have animals to look after, to have their own horse to ride, and to look after their own equipment. It gives them a certain sense of pride and independence when they can do all that.

The youngest boy came home from the Gang Ranch in BC, because my husband was sick in the fall. Hubert got real sick in March and so the kids would come over and help do chores. He stayed home until the last five days before he died, and then he went into the hospital and my youngest son came home to help calve out the cows. Brian, who lived about ten miles away, also came over to help.

That first while I made up my mind because even though I knew how to work and I knew what needed to be done, I didn't want to do it by myself. When I had somebody else to work with it was much better. I had talked to the boys and asked who wanted to come back here. When my husband got sick at first, Brian talked about coming back. His dad said we'd get him a house, but he died before that could be done. My youngest son lived with me for a while. My daughter was back and forth. I had the old Dionne house but it was cold, so I took some of the insurance money and bought myself a modular home to put on the place.

I was young enough that I couldn't afford to give it to them. I needed money to live too. I left it up to them to figure out who was going to buy it from me. Each one of them decided what they wanted to do. I think it was only fair. It was their home too.

I belonged to the Livingstone Ladies Club. We used to have strawberry teas. They started the club in the First World War when the women knitted for the soldiers. Then it became a social club. Our main function is to look after the cemetery out at Livingstone. We earn money and do funeral teas for community members and donate money to different charities in the fall.

Since I've been in town I joined the quilting club and learned how to golf a little bit. I have one son up at Swan Hills and I take some of the grandkids with me, cram as many in as I can legally do, and take the trailer and go and spend a week up there.

I always enjoyed spring and the change of seasons and the farming because the dirt smells wonderful. In town, I am on a corner lot because I could never be hemmed in by houses. I'm okay in the winter, but when it comes to spring and summer I've got to get away. At home I went outside every morning to see what the weather was like. I still catch myself opening the door and sticking my head out.

My daughter-in-law rides, chases cattle, and has an outside job. They have everything much more convenient than we did, but there's still the same amount of work to be done and things are more expensive. A few ranches might hang on but the majority will be swallowed up by larger corporations. If people can, it would be wonderful to stay with the old way of riding and looking after cattle. Everybody is so machine-oriented nowadays that they have quads and bikes. People don't really realize that they're far away from nature.

ANNE STEVICK CHECKING CATTLE AND FIXING FENCES

ANNE STEVICK

If I have to work a little harder just because I'm female, it just makes me a better person.

Always the ranch girl, Muriel Anne Stevick was born in Claresholm, Alberta on December 15, 1950. She was raised on a ranch west of the town on the eastern slopes of the Porcupine Hills with her two sisters and her brother. Her mother was Dorothy ("Dixie") Holroyd and her dad is Wes Alm. Her dreams of being a rancher came true as a result of hard work and determination. She is married to Quentin Stevick, and they raise purebred and commercial cattle, ten miles south of Pincher Creek on their Bar 15 Simmental ranch. They have a successful annual sale in December each year.

I KNEW IN THE BACK of my mind that I always wanted to ranch but we always knew that Glen would get the ranch. That's just the way it was. He was the boy and that was who would carry on the ranch.

Dad was able to teach us how satisfying it was to do a good job, and he always said to us, "I don't care what you do but always do your best." Honesty, integrity – those were probably the two most important values, and then the ethics of hard work. I always wanted to ranch but as I got older Dad would say we had to get an education. He really drilled it in. If you're a girl, unless you marry a rancher, you're not going to be a rancher. But I guess I knew somehow it would work out, that I would have a place of my own or that I'd be doing something to do with ranching.

Boy, we didn't have a social life. We thought brandings with the neighbours were the highlight of the year. Maybe six or eight families, including the Ewings and the MacLeods, who lived within ten or twelve miles were all that we knew. They used to have Christmas concerts at our little school and at Greenbank. Our family may have gone to town once or twice a month on a Saturday. We'd always have supper, do the grocery shopping, and go to a movie.

When I went to town school, I was so shy and scared that I really didn't have any friends, so all I did was study. I knew the horses liked me and the ranch kids liked me, but that was it. I finished high school in Claresholm and then I went to university in 1967 when I was sixteen. I was very shy and scared, and of course I couldn't go home. I hated it and that was when I really decided I just wanted to stay home forever. Dad was saying to us that we had to go away to get an education, but he meant not just an academic education – he also meant we had to go out and learn about life. We couldn't stay home and just ride our horses forever and we didn't know that. I actually still say my university degree was more in living that anything.

Having a degree made quite a difference in getting a summer job. I was able to get a job in Waterton Park as a park interpreter. My Granny and Grandad lived down at Waterton where he'd been a park warden. I decided I wanted to travel by my third year university. I travelled with a girlfriend around the world for about a year and a half. That was a wonderful experience in learning about life and how to take care of myself, learning that I could. I also knew there is someone up above looking after me because many times, I wouldn't be here today except for the higher power.

I was still not sure in my life, even with this travelling, where I was going to be or where my part or place was because I knew if I went home, I could not just stay home and ranch. The family farm does not have room for all of the kids, so I had to find something. I wanted to be involved. I learned that in New Zealand because I ended up on farm after farm, so I had to figure out how to get there. When I got home from that trip, I got a couple of jobs as a herdsman on a ranch up by Calgary. I ended up taking show cattle out for people.

I really think if you want to do something bad enough, you'll do it no matter what. If you have to work a little harder because you are a female, it just makes you a better person. What is really neat is to have the respect of the men in the industry. When someone comes up to me and asks my opinion about cattle or asks me to judge a show; that is really neat. They would go and ask a man quicker than they would a woman unless they really respected my opinion.

Quentin and I have been married twenty-some years now. We've been fortunate enough that neither one of us has had an off-farm job. Most of my work has involved the cattle, horses, and books. Quentin takes care of the tractor, farming. I do all the bookkeeping and the photography and catalogue work for the sale. I do help halter break and I do all of the artificial insemination. It's a pretty good split. There are times when I think it's unfair. What they don't count with the ranch wife is all the cooking, cleaning, and clothes. He did dishes once, I think. That's just sort of taken for granted, the double duties. The day our sale is over – we're getting ready for calving for the next, so it's pretty much a whole year deal. Sometimes it's really intense, sometimes it's not quite so.

I think we have a mutual respect. A lot of the cattle work is together. Our relationship is when we get upset, we yell and holler and then it's over. There are a lot of people that don't like working cattle with us because there's a lot of words and a lot of discussions, but generally we respect each other. That is so crucial to the whole operation. We don't have a lot of time to talk but when we're going somewhere in a vehicle, we do a lot of management decisions then. We definitely have the same goals and the same way of planning.

Sometimes I feel bad that we don't have children. I don't think that I could have children because I ended up with cancer. Not knowing this early on, I think that's why we never did have kids. We didn't go to fertility clinics or adoption.

I guess the frustrating part, especially in agriculture, is that there are so many things out of our control. We make sure everything that we're physically capable of doing is done, but drought, floods, and hail can wipe us out. We can deal with drought, with weather and losing calves, but then the political things happen that just slam agriculture down. That is the most heartbreaking part. But to make up for that heartbreak are all the wonderful things that happen, watching new life, watching our planned breeding calves grow into something really good, and selling to somebody that really appreciates it.

Just the fresh air, nature, admiring God's creation is I guess why outdoors is so appealing to me. I do less physical work now and it's hard for me because I love to do physical work. We have a full-time hired man. But I do consciously try to be more careful, and I don't go and clean corrals with a fork or anything like that. A person can have all the money in the world but it doesn't buy health so it is really important. We've started to realize that we have to just take it a little bit easier. As the years go by, the technical parts of our business increase, so I get to spend more time on the computer and more time doing those books. That's okay – I like that.

The bane of our existence is we don't know how to say no. When we really care about our community and we have ideas of how it could be improved, we just can't help it. We get involved, but it is a real sacrifice for people in agriculture to be in volunteer positions because either their partner has to make up the work, or the work does not get done at home. I guess about a quarter of my time is spent on volunteer stuff. I became involved with the Economic Development Board about seven or eight years ago – actually I was approached. It started with a discussion at our bull-sale about preserving the Western way of life through poetry and literature, and so cowboy poetry has become a huge thing in Pincher Creek. I feel very fortunate to have been involved in that right from the beginning. That just led to being approached to sit on the Economic Development Board as the agricultural person. That's been my interest – to bring the agricultural perspective into the Chamber of Commerce and to give them a little bit of insight as to how we look at things, and to perhaps get the rural community more involved in what is happening in the town. It's been one of the most rewarding things that I have done.

We have definite plans for retirement. We may still have some cows around but they'll be way less labour. I will spend more time doing the planning, the book part of it, and the fun parts like the photography. I still will be outside riding and walking, but it will be less heavy physical work, less stress like the night calving, worrying about our sale, and so a more relaxed way of life. It will be a different type of ranching. It'll be maybe running yearlings, not having cows to feed in the winter, being able to travel, to go south in the winter. Now after that, I imagine retirement when we get old enough so we can't really live out on the ranch.

I'm not even sure where the family farm is going to be in the future because it's becoming that they have to get bigger to have the cash flow to go on. Everything we've read, even this summer, is really scary about what's happening to the family farm. I think now there's even less of a margin in agriculture so there is even more pressure.

I think the family farm is going to become basically an acreage. That family has a nice lifestyle because they live in the country but both mom and dad go to work. They have a few cows or a few pigs or whatever. Then the other side of the coin is going to be the huge farms like the Hutterite colonies, or I don't know whether you call it corporate farming, but huge farms where mom and dad probably don't even see each other because they're running so hard and have so many employees to look after. It's really going to be interesting whether or not the family farm as it was will still be in existence, or the role of the wife of the family farm is even anywhere close to the way it was when I grew up. I hope it is, but I don't know.

CYPRESS HILLS

Although there are evergreen forests in the Cypress Hills, there are no cypress trees. French fur traders mistook the lodgepole pines for the eastern jack pine or *cyprès*, and English traders translated the name into Cypress Hills. The hills cover approximately 2,600 square kilometres, stretching through southwestern Saskatchewan and spilling into southeastern Alberta. The area around Eastend is known for its "white mud" hills, a source of clay for making pottery. It's also the home of Scotty, a fossilized *Tyrannosaurus rex*.

Points in the Cypress Hills boast the highest elevation between the Rocky Mountain foothills and Labrador, and it is believed that the top hundred metres of the hills escaped the last period of continental glaciation. In the hills, the waters divide, with southbound streams headed for the Gulf of Mexico and northern waters destined for the Saskatchewan River and Hudson's Bay.

The Cypress Hills, called "the Thunder Breeding Hills" by Indians, have provided shelter for humans and wildlife for at least seven thousand years. In the 1800's traders and white wolf hunters moved up from Montana and brought with them whiskey and smallpox. Varying reports on the 1873 Cypress Hills Massacre say thirty to sixty Assiniboine were killed. In response to this conflict, three hundred North-West Mounted Police headed west and established Fort Walsh in 1875. A few years later the Canadian Pacific Railway was completed past Maple Creek. In 1885 a disastrous fire swept through the grasslands and the forested areas of the hills. One ranch hand lost his life.

Cattle ranches, some of which carry the same family name now as they did then, became established here in the late 1800s. Many ranches still operate as they did one hundred years ago, with livestock handling done on horseback. The rolling terrain is dotted with poplar bluffs and winding creeks. Cobblestones, polished smooth, embedded in the shallow topsoil, and hiding among rich stands of native grass, discouraged cultivation in centuries past and still do today.

The Cypress Hills is a land where temperatures can be forty below one day and by the next morning a chinook will melt the snow banks until water trickles down the coulees. It is a land of intermittent drought, spectacular sunrises, red and purple sunsets – and a sky so large you can lose yourself in it.

JOAN LAWRENCE

I came to Maple Creek and I never left,
and I've been most happy all my life.

Like many local ranch women, Joan (Fleet) Lawrence came to the area as a young schoolteacher. The house where Joan lives was built along Fish Creek in 1890 by her husband's grandfather. Joan boarded there, with her future mother-in-law, when she began her teaching career in 1935. Joan is the grandmother of Christa Lawrence and the mother of Heather Beierbach.

I WAS BORN in Moose Jaw in 1915 and in 1918 my father died of the flu. He was the CPR express man between Moose Jaw and Calgary. My mother was left with two little children, one almost three and one ten months – I guess she was very lonely. Since my grandparents from England decided to go homesteading in Manitoba, my mother decided to follow them. She was later married to a farmer there, and I had two stepsisters, making four girls in all.

I spent a lot of time in Neepawa, Manitoba, going to school until there was a school built at the homestead. I did go to the rural school for grade three. It was Birdina School, on my

JOAN LAWRENCE WITH TURKEY FLOCK IN THE 1940s

stepfather's property, and named after his niece. I remember I was the janitor. I couldn't have been too old, but I got ten cents a night for sweeping the floor, keeping the boards clean, and maybe laying in the wood for the fire. I don't remember lighting the fire. I ended up with fourteen dollars at the end of the school year, and that was a fortune! I finished my high school in Neepawa, then went to Normal School in Winnipeg. At seventeen, I had trouble getting in because I was young. I graduated when I was eighteen. My parents certainly encouraged me in my vocation as a teacher. I remember my stepfather driving me with the horse and buggy to interview people when I was seeking a school. And I wasn't very big and I was young, and when you have about forty pupils, I could see they'd think, "She couldn't manage!" So that's how I didn't get a job in Manitoba. Probably a man would have likely suited their needs better.

I went to live with my aunt and uncle in Regina. After Christmas, I got a Saskatchewan teacher's permit and applied for a position. I was offered three schools, and only one could pay any money. So, of course, I took that one, which was Smithfield School at Maple Creek. I received thirty dollars a month, but fifteen dollars of that paycheque went towards my board. I boarded here, where I've always lived. Mrs. Lawrence was a widow lady and Russell was here.

Russell and I were married in 1935. Well, in the next two years I had two children and I wasn't able to teach, but I did go back and teach for another couple of years, and get my permanent certificate. I guess that's what I was aiming for. My husband wasn't all for my going back to work after marriage because married women did not work those years. I had some rebuffs from neighbours – they just didn't think it was right. Anyway, I did it.

I was surprised, a bit, when I became a ranch wife – the lifestyle was different. There were conveniences in this house that I never thought you had in the country, like bathroom, running water. I never knew what it was to not have running water here. It was a manual pressure thing. Yes, that impressed me. And a car, we never had a car – my folks, lots of people didn't, just horses.

I was very busy because there was always so much help. A lot of help were drifters, they just came for their keep. I was busy with gardening, taking care of chickens, baking, sewing, quilting, making mats, and knitting. I baked bread about three times a week because we did have sheep after that, and two sheep camps. So you're forever baking bread. The ovens are full of it. But I rather liked that because I always liked to cook and I didn't have an aversion to cleaning that people have now – I'm forever at it! But that, with children and keeping up with housework – oh, I was always trying to do a bit of painting or papering in the house. Russell's mother had been here and she had hired help, but she wasn't able to do so many things, so I pitched in as best I could.

We had five children and four of them are involved in ranching. Heather Beierbach, Frances Needham, and Eric Lawrence ranch in this area, and Donna Hanel at Swift Current. Elaine Isabelle lives in town. I have fourteen grandchildren, eleven great-grandchildren, and two great-great-grandchildren. The best part of raising children on a ranch is their sharing in the work and the decision-making. I think that's wonderful. In the summertime they played along the creek, and they always had playhouses and spent hours in these playhouses, which children don't seem to do anymore. We had a room upstairs that wasn't finished – they played school, painted on the walls, the old plaster, crayoned or chalked or whatever. They never seemed bored. I never seemed to have to think of something for them to do. They didn't have a lot of toys. I think they were happy, happier than today's children. That is urban children, anyway. They got involved with riding and chores, irrigation, haying, and feeding. Well, of course, Eric, it was just natural for Eric to do all these things. He helped with everything. He couldn't get through school fast enough! The girls all did their part in helping with the ranch work, too.

The children were all born in the Maple Creek Hospital. It was a long way. You just jump in your car and you're there in ten minutes now, but it wasn't always that way. And, of course, cars weren't as good, either, and the weather played a big part. I breastfed all my babies, except Eric. He was a preemie, in an incubator, and they felt they made a mistake by not using mother's milk, but that's the way it was then. Russell's mother had several of her children at home. I know Russell was born in this house. I think that generation, they all pretty well did. They often had a midwife come. Russell's mother delivered babies.

At first, I didn't get to church regularly, we didn't have too great a road here. Weather conditions controlled what you were going to do – you didn't just start out. So I wasn't able to attend like I do now, but as the children grew older, we wanted to get them to Sunday School. You made more effort to get in. I am quite involved with the church and all the activities there – vestry, warden, Altar Guild, ACW, Pastoral Care, and I have led youth groups. Now I don't drive on the highway, but that's my own decision not to drive on it. I still drive the eight miles into town, and I drive to Heather's or Frances's. I would hate to lose that – I don't think I could stay on the ranch if I weren't able to get myself around and go to do the things I want to do.

In 1960, 20th Century Fox, the film company, were here for eight days filming *The Canadians*. We had over one hundred head of horses brought in for the horses that were to come off these cliffs in the movie. My husband was a stand-in, he was an Indian. That was the word then, you didn't say Native. He had a costume, you know, and he had to be on the set at seven in the morning.

There was a catering service from Regina that brought their dinners. They stayed at the park, the rest of the people, but those with horses stayed here. I fed twelve here for supper. We had the RCMP who rode, who were stand-ins for the ones in the picture. And I had the wranglers, the ones who looked after the horses.

Robert Ryan and Teresa Stratus starred in it. But there were so many local people involved. It was most thrilling. I think, for some of these older ranchers, that it was the thrill of a lifetime. It was something different and very interesting – those that gauged the light and the rays of the sun for picturetaking, and the camouflage for the telephone poles, there was a lot of preparation. There was only one half-day in October that they couldn't work. The Old Timers have had some reruns of the film, and I've never missed it!

John Lawrence Sr. came here in 1886 after his son, John Lawrence Jr. had been in the fire in the Cypress Hills in 1885. Here he almost lost his life. His hired man, Mr. Parks, did lose his life. John lost his wagon and team of horses, really his possessions, you know. But he managed to roll up in a blanket in a swampy place up in the park, and he wasn't burned, whereas his cousin, Gordon Quick, was badly burned – he, too had rolled up in a blanket. I often wondered, when I first saw this man – he had no earlobes, that's how burned he was. Anyway, they survived the fire and when John Lawrence Sr. came, he was really taken with the country, so he filed on this homestead. Right here. They moved up in '87 and lived with John Lawrence Jr. for a little while, and built a log building, which they lived in while this old house was being built. Then Mr. Lawrence Sr. took sick and died in 1902, but he'd built a sixteen-room house in town for his wife, who was Scottish and not really happy here. So the "Castle" house was made with verandas and balconies and that. I regret that Russell and I didn't try harder to save it. Fight the cause a little bit. It was demolished.

Russell was born here in 1903 and had the ranch until 1982 – then it was turned over to Eric. So, our generation ... Eric doesn't have a son, but his daughter – ranch life is her lifestyle, she knows what's going on from one end to the other. I rather think that she'll be the next generation to take over. And I think she'll be quite capable doing it. I think maybe they're guiding her along the way, and we'll see what the future holds. It is kind of fascinating. It is the only ranch in the area that has passed from son to son for four generations – it was never owned by anyone else.

I'm still very interested in everything that's going on and they keep me posted quite well, and I appreciate that. Christa came over last night and told me what they were doing today and what they'd done while I was away. The children take an active part in ranching. They all work and share in the riding, feeding, whatever. I think they enjoy being a part of ranch life. I think they're very fortunate to have this, to share in what's going on.

I guess I just expected to retire here. It's what Russell wanted, what Eric wants, and I think maybe it's been helpful to them. I always babysat – the children came over here when they were little, and it was so wonderful to have those little people. They looked forward to the morning visit, while the feeding was going on.

I love the yard and garden and picking berries and things like that. I still keep a few chickens and I still do the yard myself. I think the time's coming when I can't mow and clip hedges and so on. But so far I've been able to do it. I do have a few head of cattle and I did have thoroughbred horses, because I raced after Russell died – we still had racehorses and they did race in the Calgary-Edmonton circuit and the Regina-

Saskatoon circuit. I think that's the most thrilling experience one can ever have, being in that winner's circle. You're just so geared up. We had wins with Ready Roughneck and Spinner and Homespin in the 1980s. I did enjoy it. Then we quit racing and I have one left. Of course, I still belong to the Thoroughbred Horse Society and get all the magazines and follow up. But things are changing there, too, like everywhere else.

So, there are things you have to cease to do. I kept bees and I quit that, the work was too heavy for me, and you realize you've got to start cutting down. It's hard. I think I've been happy; I haven't been yearning for changes. I just hope I can stay here. But you know, should I not be able to drive or care for myself, well, you have to accept that if it ever comes. I wouldn't be able to change that; I'd have to make the changes, and I hope, graciously.

HEATHER BEIERBACH

My photography tells Roger's message about the communion of spirits between man and animal and shows the beauty of that.

Heather Beierbach, Joan Lawrence's daughter, was a provincial director of the Saskatchewan Stock Growers Association. She ranches fifty miles from Maple Creek on "the dry side" of the Cypress Hills. Her husband, Roger, died accidentally five years ago, leaving two children in their twenties, John and Eve, and a forty-five-year-old widow. Heather's creative photographs, dedicated to the genuine artistry evident in ranching, have helped her to cope with her loss. Her calendars became a memorial to her husband.

I BEGAN TAKING PICTURES because I was always riding and helping my husband, and he took great pride in his horses and steers, and how he handled cattle. He studied it and took clinics – it was a passion of his. When you're in on that sort of thing, the passion rubs off on you and you start to see those things and understand the difference between a horse's head being thrown in the air and one who's quiet and gentle and isn't being yanked on. Those things become beautiful and when you see them, you think, "Oh, I wish I could capture

HEATHER BEIERBACH WITH HER FAMILY –
JOHN, HEATHER, ROGER, AND EVE

that." So that's literally what I did – I began carrying a camera, not with any plans for it at all. Then some people who were doing a calendar happened to go through my photo album and got pulling pictures and asking for negatives, and no, I didn't have them. I had tossed the negatives! So, I've learned the hard way on a lot of things. Every time I got pictures back, Roger would say, "Oh, look at this one, look at this one." He was excited about them and he was excited about me doing it, and he really kept pushing me to carry on and to do more – to do something with them, not just to send them here and there.

I made some cards; I got that much done, then he wanted me to do a calendar. I finally got at that. He and my daughter had been pushing me really hard on this. In fact, Eve called me in January, this was 1998, she at that time was in Australia, "Have you started on your calendar yet?" and I'm really hard to push. I said, "I got my cards done without you prodding me last year. Leave me alone and I'll get it done." Well, I didn't and I believe things are written down in time, I guess, predestined. I believe that everything happens for a reason. And I had not started my calendar until July, and then Roger – I had already taken the pictures to the printer – and then he was killed August 1st. So it was already in motion. There again, the small printing plant I had taken them to, in Medicine Hat, had shut down – the whole staff was on holiday. And that gave me time, and in my daze, like the dazed state you're in afterwards – I think when you're in that deep of grief you almost go on like a robot – I decided that no, I can't do this, but I thought, "No, now I must keep it."

So I did, and, I have carried on for the same reason. My photography tells his message about the communion of spirits between man and animal and shows the beauty of that. I don't know how far that message goes; I don't know how much impact it has. But he did feel that it was important that people beyond our way of life understand that we do have feelings for our animals, that we do care for our animals. I think it is an important message, so I'm still doing it. It's a memorial for that, and I hope it influences things too. I feel that Roger in his life was a huge influence – I saw lots of kids who emulated him. Instead of it being a square thing to be a cowboy, they were proud of it – not that they were better cowboys, but they were proud of who they were and what they were. Instead of being bent by peer pressure some of the kids actually ended up being leaders. Roger always talked to them; he always had time for them. In his family, he wasn't encouraged; his father often put him down. Roger was gregarious, he didn't go by all the rules. He was always full of vim and vinegar, and so you can see that happening. He didn't get encouragement, so he gave encouragement to these other kids, and to his own family. As wonderful as my dad was, he never really ever gave me a pat on the back, and if he did, I would catch him doing it – it wouldn't be in front of me. But, Roger – if a kid was trying to learn how to rope, he was so helpful and any trying was rewarded. And he was the same with animals, their best try was just great by him and he always took time for all of them.

Roger took classes and went to clinics with Bill Collins, for instance, who was a Canadian calf-roping champion and also a cutting horse champion. I think he really absorbed things well, and took it all home with him and lived it afterwards – he didn't go and forget what he'd seen. He kept going back, and he read and read. I don't think he ever had read when he went to school. He bought books and read, he was an avid reader of anything to do with horses and history and cowboy lore. If I read, it's often fiction, and he'd say, "I don't know how you can waste your time on that." And here was this man who never had finished school and had never been a good student!

Both the children were at the ranch when Roger died, and there again, I really think that things are all planned, and we just play the role. They had both gone off to school, John had taken a farrier's course. I tried to encourage him to do more things, other things away from home, but he resisted and did not want to, and we don't push John, either! You might influ ence, you might talk to him, but ultimately it's going to be his choice. He certainly knows his own mind. Eve had taken animal health technology, and she'd come home and worked for a while at the animal clinic and worked at home. She went to High River to work at a feedlot, then she and a girlfriend went to Australia. She was very homesick while she was there. When she came back, she came back to stay at home. That is where she ultimately always wanted to be. So when Roger was killed, both of them were home, both of them had made that choice to be there, and John said to me, "The one thing about it, if Dad had to go, it was the way he should go." It was sudden – he never had a moment's realization of anything having had happened to him. Being hit by lightning was so right for him. People say God doesn't do these things and when I think about that, I think, "Oh, yes He does." For whatever reason, I'm not going to say that I think it's right, wrong, or in between, I didn't like it and it hasn't been easy, it's been a bitch, but He did – He gave him that gift. And, as John said, another thing, we were all happy. There had not been any last-minute words said in anger about anything. Everyone was totally content at that time. You're kind of pulling at straws to call that a gift, but it is.

There aren't any routine days on the ranch as such – it depends on the season and the jobs at hand. Now that Heidi's there, she goes with John to feed, and also, since Roger is gone, I try to stay out of my kids' way. I try to be there to help, but I try not to be there to push and drive and organize – I try to let them do that. Haying season, of course, there's more routine and we were all involved. The kids learned to run machinery young. I learned to run all the machinery as well. In the winter, feeding time was more routine, again, because everything was fed every day. I would go and help Roger, and when the kids were older and at home, they would feed as well. So if John was feeding cows, for example, he'd run the tractor and I'd go with Roger with the team. Earlier, when they were little kids, in school, often I would feed with my own team and Roger would go his direction with his own team or else with the tractor. We were all integral in the work. When the kids were home on the weekends they were out with us, doing the feeding and choring. Sometimes, if it was a big project, they would miss school. They got to be very proficient, I would say, horsemen, but one is a horsewoman.

We do use horses a lot. Anything that had legs was either ridden or driven and sometimes both. Roger grew up, actually, with goats and he drove these goats when he was a little boy. I still have the little wagon his great-grandmother had had built for him, and his uncle, Howard Buchanan, had made a little harness for these goats. He even drove them in the Medicine Hat parade. I had never had goats, but Roger got goats when we were first married and I was the one that fell in love with them, quite seriously, afterwards and there are still goats on the ranch – I can't imagine having a place without goats! I get a lot of pleasure out of them, especially when they are kidding. The kids are the most delightful little devils in the world, they just light your soul and light your day. There is just nothing cuter and more mischievous than kids.

John had a mule that he rode as a little boy, we had bought her. He got along with her so well, but later on Roger ended up being the mule man. The mules loved him and he loved them. Mules are quite different than horses. One horse trainer said, "A mule is like a horse, only more so." My description of them is that they are very eccentric and Roger would always say you have to be smarter than the mule, and that took some doing. They are a very fascinating animal and we still have probably about ten head that are broke to drive. Roger sold mostly foals, but he did sell some teams as well. Also, we raised draft horses, Percheron horses. There's something awesome about a draft horse and I love them. I remember feeding with Roger one day and John was feeding with the tractor and I said, "That must be awful for him, driving that tractor every day. Maybe he'd like to feed with the team sometimes, too." And Roger said, "I never thought of that." To us, he was being robbed of the pleasure. Lots of times we would use different teams through a winter and start young horses and there was such satisfaction watching a young horse develop and mature and improve. You didn't want to drive a young two-year-old for too long because it was too much stress on them, but for a little while, a few weeks, then that was enough for that year. Next year they were ready to be used some more, and the mental maturity would be there, they were ready to go then. It was fulfilling to see them develop and learn and get more secure every day, and more relaxed and understand what was being asked of them. We still use horses, though the kids don't use as many. They did right away because they thought they were supposed to, but we had trouble enjoying our work after the reason we enjoyed it, to begin with, was taken. The pleasure is coming back, but it takes a long time.

I finished my grade twelve and I went on to university – again my mom encouraged me to. I was seeing my husband at the time, and I wasn't that sure that I was really that interested in going to university. I quite enjoyed school, and I did enjoy university, but I was terribly lonely, never felt that I fit in, and I think it takes a little maturity to feel that you fit in – I didn't get that until much, much later, where I was comfortable anywhere. I took two years of education and then got married, and I'm not sorry for a minute! School was very important, I guess, because my mom had been a schoolteacher and education was very important to her, and it was also very important to my dad. He had a very limited education, he actually rode, when he was a boy, to Maple Creek with his younger brother. He was kept at home, I believe until he was eight. If you can imagine, sending your children out on a horse eight miles to school! Even I wouldn't dream about doing that with my kids, who are more independent than most, but we maybe overprotect our children. Anyway, that was what they had to do and they did it. As a result he didn't have that much schooling, but he was extremely well read.

When I was growing up, Mom wanted me to be involved in things. She wanted me to take piano lessons, and when I didn't want to, she said, "Well, if I had had those opportunities." So of course she guilt-tripped me into it, and later in time, although I have absolutely zero talent, I do have a huge appreciation for music, and I did learn to enjoy it. So there was something gained from that. I'm glad I did it. Occasionally, I play at church. I don't mind helping out when they need someone because for one thing, it gets me there, and I feel that I have a church family as well. I always think of my dad, he never pulled any punches. There was a lady who played at church, I probably don't even play as well, but anyway, she always said, "I never wanted to play, but my dad always made me practice!" And, my dad always said, "Gee, I wish to hell he hadn't!" If I hit a wrong note, or fumble-dumble across the keyboard, I remember that, and wonder, "Oh, no! Who's thinking that now?"

I guess I'm interested in everything and that's a bit of my undoing. I have too many interests, too many curiosities, too many things that, "Oh, I would love to do that." Especially things with my hands, and I guess they're creative things. I remember as a child, somebody talking about Ken Wright tooling leather, and Ken is still carving leather and he is still a master and I've done some too, but I only wish I could reach that level of achievement and craftsmanship. I've taken a pottery class, I'm still intrigued by that, but I'm mostly buying it, I'm not doing any of it, but it's something else that captured me. Chaps are really what I do the most of, and that started when Roger was making me a pair for Christmas and I got helping him and one thing led to the next. And a horse trainer, Bill Collins, showed me how to measure people and I pretty much design my own patterns now. I've made over a hundred and fifty pair. For a real chap maker, that's not so many, but when that's fitting it in with the rest of your life, that's quite a few.

The children are ranching. I did try at different times to give them direction to look outside the box. At one point there was the Critical Wildlife Habitat, the name has been changed but that's what we all still know it as, that sounded like quite a threat. We have mostly Crown land, most people in our area do, and basically it's quite low-production land, quite arid. So that was a threat to our livelihood and it actually still is. As you're not selling anything and as long as nobody else wants it, you kind of coast along at this present time. It's really too bad that John hadn't pursued something, like becoming a veterinarian. He could have returned to this life, there was still time. Well, there wasn't time, the way it turned out, but there still could have been time, and he could have afforded to live the life he actually enjoys. They grew up doing absolutely everything with us and ultimately they must have got the same fulfillment from it, and I guess that contentment and satisfaction is in your soul. That must be what we do this for, because God knows, it's not the economics, is it? My encouragement didn't go very far. And Eve still wants to do this. She's married and she and her husband, we bought a place close by and they're trying to get started. I wonder just what the sanity was there, too, but it was what they wanted.

Roger enjoyed his family – he wanted his kids and myself with him constantly. He expected we'd be capable of doing anything he could do, and he taught us how to do everything he could do. For me, I'm not an athletic person at all, still we worked at it and he encouraged. He was always very positive and he encouraged us in everything we did and he taught me how to rope, just calves, but I can handle a rope on a large animal if somebody else has it caught. And you know, where to be and how to be safe and how to handle things. John and Eve learned all these things at a very young age, but he was such a good teacher. We were so prepared, and it almost seemed, when he was gone, when he was taken, we already knew how to do everything. Mind you, Eve and John weren't children anymore, they were young adults, but they'd been doing all this work from the time they were little children. I don't remember when they couldn't ride – they were always with us for everything. That's another reason why there was such a void when he was gone, because we shared all of our work. Afterwards, even when we had a good day, I'd come into the house and I'd want to share it with him, I wanted to tell him about what we'd done, but he wasn't there to tell.

I often refer to him as the hub within the wheel of our family and we were the spokes and he had us all circling around him and usually sending us off in different directions. He didn't mind telling people what to do, he didn't feel guilty about that, no qualms at all! I often think that the way he lived his life and the lessons that he gave us all were because somehow, even without knowing it, he knew that he was going to be gone from us. Actually, it was a premonition I had from very early on, even before we were married, but mostly when we lost our baby son – then after that I felt that he would be taken from me. I had that premonition, I had shared it with a friend, and when I did lose Roger, he remembered that. I had only shared that with one person. But it was something that really did haunt me for a lot of years of our marriage. Then I started to relax. Roger was fearless, not careless, and there is a huge difference. I was looking forward to our growing old together because he was always fun and he was always vibrant, but I never got to do that.

As for the future of ranching, it's frightening. I don't know, I am really quite scared. I see all kinds of things infringing on our lives. I see a lack of respect for our livelihood and a misunderstanding by urbanites and some left-wing groups. I think we will have to address that and make people understand what it is we do, and why we do it, and it's, in the ranching end of things, raising beef. It's not unhealthy, it is a healthy food, and we have so much work ahead of us. Not only that, our costs are so astronomical and our product doesn't seem to be rising at all in value compared to costs. I don't know how we can afford to do this anymore. We're still hanging in there because that's what kind of people we are, we have strength of character and we know how to stretch things and make things still work. But ultimately it's a lot to ask of anyone else to come into this life, or to stay in this life. We can all compare to professional jobs in the city – teachers, nurses, doctors, lawyers – I mean they make huge salaries and have access to money which we don't have. We are worth a lot but we don't actually have liquid funds. So I don't think there is anything else that compares to our way of life for fulfilment and contentment. However, there is also huge frustration and huge stress. You have to let us do it. I hope to do it, I hope we can always do it, and I hope that it is always appreciated. We really do have to promote, not our lifestyle, but our product. Do whatever we can to popularize that with the consumer, so that they will want to eat it, so we can afford to raise it.

HEIDI BEIERBACH ON RIDGE RUNNER, 2002

HEIDI BEIERBACH

I could be out making fifty thousand dollars a year
and I just love this life way more.

Heidi Beierbach's parents, the Moorheads, homesteaded on Piapot Creek southeast of Maple Creek in 1889. Several Moorhead families now ranch along the creek, and the Moorhead School, where many of the earlier Moorhead children got their basic education, still has chalkboards good enough to use. Whenever the Moorhead name is mentioned in the area, stories of cattle, horses, rodeo, music, and dancing come readily to mind. Heidi Beierbach, the eldest child of Laurie and Melvin Moorhead, graduated from Maple Creek High School in 1994. Heidi is married to Heather Beierbach's son, John.

I HAVE ONE BROTHER, Clay, two years younger, and one sister, Shauna, nine years younger. When I was little I remember being out riding horses and helping Dad with the sheep and calving. In the winter we went behind the tractor on the toboggan and if ever there was a cow having problems, we were right down there, in his way, I'm sure, but we wouldn't miss out on that.

We went on a wagon train in 1984 from Elkwater to Murraydale and I remember having so much fun. I kind of wish we could go on one again. My mom was pregnant and she still rode along and we had the covered wagon and my brother rode the little Shetland pony – I don't know how he made it that far on that thing. We were going to the Murraydale Stampede and we stopped off at Fort Walsh. A bunch of the Moorheads were there, and my Grandpa Hobbs, and they had their wagons, too, because I think Grandpa Hobbs was one of the teamsters. There were a bunch of us cousins and I remember the campfire because my Uncle Marvin had his banjo – there were three of them that played music until all hours of the night and us kids, we just ran rampant around there.

Dad made us a little stone boat, and my brother and I had to harness our little Shetland pony, but Shetlands kind of get stubborn. But we took her in the parade at Piapot a couple of times. I remember I was dressed in my little outfit from the wagon train with my little hat.

I'm really glad I was raised on a ranch. It's an excellent place to bring up kids. I just think the kids in the city don't know anything about prairie oysters and wide-open spaces and gophers, and they don't know where milk comes from. I mean some of them do, I'm not saying they all don't, but lots of them have never experienced it, and it's a totally different way to be brought up. I wouldn't have wished anything else – a small school where everybody knows everybody and you can send your kid home to the birthday party on the bus and not have to worry about them. Out on the ranch I think kids learn not to be afraid of animals. I remember being a little girl riding our pony or even a horse and, you know, you just jump on. You have trust in your horse or you give your kid a horse you know you can trust. And kids learn to be responsible.

I was a 4-H beef club member for nine years. 4-H was really good for me, with leadership, and I've got close friends that I met at 4-H camps. I'm still involved, as I've marked judging cards at Regional Fair and record books for Bear Creek Club. Someday I'd like to be more involved with it, actually. My kids will be in it, for sure, if I have any.

I started out at the old Jasper School in Maple Creek, then to Sidney Street School, and finally high school. From there I went to the University of Saskatchewan. I was going to take pre-vet, but once I got there I realized that it was an awful lot more work than I'd anticipated and I ended up taking my degree in agriculture with an animal science major and an ag business minor. And during that time I worked several summers. I was in a co-op program, so I ended up on a dairy farm for one summer.

We milked a hundred head three times a day, so I was up at five o'clock in the morning. It was a family of five that owned the dairy, and whatever milkings they didn't do, I did. I usually worked an eight-hour day, but silaging and that sort of stuff was longer hours. I also learned how to drive the square baler and all the perks of fixing that. And the stacker – to manoeuvre one of those things, that's quite a little task, but I got onto it.

And then I worked at a pig barn for two summers. I did mostly farrowing. I kind of like the little babies best. Now I can say the only thing I haven't worked with is chickens, but Dad used to set the incubator so I remember having baby chicks at home.

Then I went to Australia for six months in 1998, and worked on a cattle ranch, so that was another new experience because they don't have a winter. They had purebred Maine Anjou, not a common breed in Australia but they had imported the semen from France. I had to leave before they showed their cattle in a big show in Perth. And their weaning practices over there – everything just runs together. So they had this idea to stick a piece of wood in their noses so the calves wouldn't suck the cows. It worked at first but then it was a disaster. They all fell out because they were too small.

I had to come back for my last year of university. My thesis was on interactions between cattle and deer, antelope, that sort of thing – how they interact out on the prairie and if they eat the same grasses and if, say, antelope are wrecking pastureland that cows could be on or if they can co-exist on the same rangeland.

When I was done university I had a one-year contract at the Swine Research Centre but I actually decided in about December that I'd had enough of this pig farm and I wanted to come home. I thought, I'll just come home and help Dad calve out his cows, because it was spring, and then I got back here and the vet clinic was looking for somebody for a few days or weeks and I thought, well, I'll just do that. Then I met John and I got married and now I'm still here. So, I came home to calve cows and I never left.

When John and I were dating, I was going to go to his branding. So I said to him, "Like, I can ride one of your horses?" And he says, "Oh, yeah." And he's got this – he calls him the black horse, and he's this special horse, nobody gets to ride him, but I got to ride the black horse so that was good. I was out there riding and everything was going good and there's quite a few people that show up there. I knew most of the people in the community, but I was kind of, not on display but, you know, they were trying to see if I was going to cut it out there on the ranch. And so we get just to the branding corrals and this calf ducks back, and I went to get it and the black horse decided to start bucking in front of all these people – I landed on the ground and actually he wrecked my back a bit. But it's nothing that can't be fixed and I still carry on, but it was quite the initiation to the branding. But, of course, once I got into the branding corral, I threw calves down. I wasn't going to let anybody know I was in a bit of pain. So I made the cut, I guess.

Out there on the ranch, there's Heather, John's mother, and John and I, so it's a family affair. We live in a house just up the hill from Heather. I've been married two years and I am working at the Animal Clinic, a veterinary clinic, two days a week, so that keeps me busy – gets me to town for groceries and I get to interact with people, so that's good and it's a little extra income. I try to help out every day when I'm not at work in town. John seems to have a hard time keeping me in the house to do the cooking and cleaning. My dad had me involved with his ranch for so long that it just doesn't feel right if I'm not out helping.

Beierbachs have a larger land base and it's wider open spaces than at Dad's, but we did stuff on horseback at home, too. We were fifteen minutes from town, so if you needed a loaf of bread or a jug of milk you could just run to town. Now we're fifty miles from Maple Creek or thirty-five from Consul, so whatever you need, you have to get when in town because it could be awhile before you get back. Everybody wondered how I'd get accustomed to being that far from anywhere, but I always told Dad I was going to become a hermit and live up in the hills, up at my Grandpa's place. When I said, "an old spinster," he said, "Heidi, don't ever say that because that's what you'll become." But I just wanted to have my own cows and my own little place and now I'm happy out there with John and his mom.

I could be out making fifty thousand dollars a year, and I just love this life way more. There is just nothing better than seeing a newborn calf, lamb, or foal trying to get up for the first time, or its momma licking it off. I wouldn't change places with anyone. And I'm still using my animal science degree, ag business, to some extent. I have the degree hanging on my wall so it can't ever be taken away from me.

I love working alongside my husband. Some days we get along better than others, but that's what it's all about. There aren't many married couples who have the privilege of working alongside each other every day, and we rarely go anywhere one without the other. I can swing a hammer or drive a tractor along with the best of them. During spring I spend a lot of time on my horse checking cows during calving season and we go to several brandings. In summer I spend quite a few hours on the swather, cutting hay. Dad and John must both think I might wreck the baler, as neither one has let me operate that. In the fall I help John get colts ready for the Cypress Hills Registered Horse Breeders' sale, and then there's weaning. There's always some fencing to do, and in winter we have cows to feed. John feeds with a team sometimes, and there's pitching square bales on the hayrack and off again. John wants to cut some colts right away, so that'll be interesting. Dad used to cut a few colts on his own, too, and I helped with that one time and, oh, I thought that I was going to get my head kicked in, so I don't know how I'm going to do.

In early January when it's colder, we are halter breaking the baby colts, and that's always lots of fun. At Dad's we had colts but I never really had much to do with them. John has taught me lots because in the beginning I was scared. And those little guys, you've got to get right in there, but then I realized that they're scared, too, and they just want you to help out. And John's breaking a few colts to ride, so I've been there for that. He usually gets on a couple of times before I get the nerve to help out with that.

It must have been about 1986 – John's dad had some mules and John decided to get a jack, so he has mules, and Eve and Travis – that's John's sister – they use the mules to feed the yearlings in the wintertime. John had a wreck with the mules before I met him and he broke his ankle. We still have a bunch and he halter breaks them, but I don't think he does as much with them as his mother or I would like. But our jack died this spring and I asked him if we could get another one. When I first went out there, I was scared of them because I've always heard mules can kick and they're nothing like a horse. I still don't know all their little quirks but I've learned lots about them and I kind of like the little babies. And he told me that he's not sure he's going to get another jack. I'm going to have to work on that. They have a bunch of Percherons, too, so they use those to drive, too, to feed in the wintertime. I can harness the mules but not the big Percherons. I told him he's going to have to get me a stool, or we're going to have to rig something up with a pulley on the ceiling so I can harness them myself.

At Dad's there were always the baby lambs, and the sheep were mine and Shauna's, mostly because Dad had had them for quite a few years and he wanted to get rid of them. Shauna and I always said, "You can't get rid of them," so finally he said, "Well okay, if we're having sheep, then you guys are looking after them." So we had to get up and feed them. He usually did a couple of the middle of night shifts because sheep, you kind of have to check a little bit more often than your cows, and we used to lamb the end of February, the beginning of March, when it was usually about thirty below so it wasn't much fun to go out there and check.

Dad always liked to tag the calves just as soon as they were born, and most of his cows were pretty tame because he has Angus and Hereford, and crosses the two. But then there's a few of those old black cows that I wouldn't get off my horse, and if you did – I remember a few times we had to get the truck down there and Dad would say, "Okay, go now," so that the truck was in between the cow and the calf – she was still running around, and you'd get done and back in the truck just in time, so that was always exciting.

Roping is something I've always wanted to do, and at Beierbach's they dehorn the steers in the spring so I'm learning to rope. John taught me how to swing a rope properly and

Heather's horses – if it weren't for Heather's horses, there's no way I could have ever learned, as they put up with pretty much anything. This spring John sent me out to get a starving calf from the heifers, so he said, "Just drive out there and you'll be able to pick it up." So I went out there. Well, there was no way I was going to catch him. He was running around. I had the horse in – I thought, well, you know I've been roping for a while now, I should be able to maybe catch him. Well, I ran around and around. That calf had more energy. I don't really think he was starving, so finally I gave up. John had to get it later. I guess that just goes to show I'm not ready to be out of the pen to rope yet.

I was first approached to enter the ranch rodeo three years ago, with the Moorhead team – they've been doing this since the ranch rodeo started, which is sixteen years. It's my cousin's spot [Derek Moorhead] I've been taking ever since. With the team penning, all five members of the team have to be in on that – we're given a certain number and you have to get three of the animals with that number out, so that's usually one of my jobs. Three riders go into the herd and the other two stay at the line because any cattle that don't have your number on can't cross the line – so they're the holders. You have to get the cattle into the pen and get the gate closed, and the fastest team wins. The first year I was in there we won, so I told them that must have been me, but last year we didn't do so well so I can't take the credit for that. And you have to learn certain rules. You have to wear your chinks or chaps, hats, and western attire and actually be on horseback. Another event is the branding. There's a pen of calves and one member has to go in and heel a calf and drag it out, not be rough with the calf, and you have to catch two hind feet – one foot gets a time penalty – and drag the calf to a circle. Then you have your two wrestlers who wrestle it and you have a person standing at a can of paint with a branding iron in it. That person has to run over to the calf, put a legible brand on, and have his iron back "in the fire." The fastest time wins. I have been the brander and a wrestler.

They have the doctoring, where there are two ropers and they have to head and heel a dogie and another team member rides in, gets off his horse, slaps a back tag on the animal, and gets the ropes off. And of course you have to rope two feet there, too. I think you're docked marks if you just have one foot. I'm usually a watcher on that one.

Wild cow milking is next and that's a free for all. One person ropes the cow any way possible, but you've got to make sure that the other two team members can actually get to the milk. The first year, I got roped into that. Well, they had the most wild cows and my brother, who was the other one on foot, the muggers, they call them, told me, "Stay with me and don't lose your roper." Travis Portz was roping and I'm out there running around and these cows chased. All of a sudden, I turned around and it was too late. I was down on the ground and I squished in the tightest ball because I had my brand new hat on and I didn't want her to step on it, and this cow just literally jumped right over me. I got up and I was a little dirty but I wasn't going to quit. I was still sticking with my men, and this other cow came to get me – that was the closest I've come to death. Travis never did get a cow roped so we never did get any milk, but I went in it again last year – this time I'll stick to the fence and I'll wait and I'm just going to run. So I'm usually in on that one.

Then they have the wild horse race. Every team takes in one of their saddle horses and they let them all go – you have to catch your rider's horse and saddle and ride to the end as fast as you can. I'm usually a spectator in that one. And then the bronc riding is just a free for all, and usually our team dresses up our bronc rider. He usually puts on a little bit of a show. He doesn't usually stay on the whole time, but at least he's a little entertainment for the crowd. So I'm looking forward to that again – lots of fun.

I'm a member of the Merryflat Ladies' Club. Merryflat is actually an old school house where we have our Christmas pageant and party and showers for anyone in the district getting married. A bunch of ranch ladies get together once a month except when it's summer or busy, and it's our day to get out and socialize. We also have progressive suppers. We usually start at one person's house for appetizers, move to the next person's house for soup or salad, and then go to somebody else's house for the main course. Then we end up at the school for dessert and a poker game. The guys have a big poker game. A few of us women have our own little poker game, but I always lose money so I figure that means I shouldn't be a poker player. But it's always lots of fun.

I don't know what's going to happen with this BSE or with our industry. I think it would depend on whether the United States realizes it was only one cow and if they're actually going to open their borders or if it's just going to remain political. Prices are not going to be what they were and there's so many people, I think, that don't understand – even in Western Canada – this is how we, as ranchers, make our money. We have lots of people that are supporting our industry, but we have some retailers selling American beef and beef from other countries, and I wish people would just stop going to them.

You're never going to get rid of ranch women. You have to be strong-willed and love what you're doing. It's a day-to-day thing. As long as our industry will let us – like look at all of them. I'm only a small part and some of these women do everything. I mean, they're coming on strong. All the young girls you see that are out there roping just amaze me. They just want to be right in there. It's great to see, so long as we don't lose touch with everything – as long as this BSE doesn't get us to the ground.

CHRISTA LAWRENCE

You're never bored here, and if you are, your dad'll find you something to do.

Christa Lawrence, a sixteen-year-old high school student, lives on the Lawrence Ranch, southwest of Maple Creek, Saskatchewan. She is actively involved in this family ranch, and would like to continue to ranch in the future. Christa is a fifth-generation rancher on a family ranch that, in 1999, received the Environmental Stewardship Award (TESA) from the Saskatchewan Stock Growers Association.

MY DAD SURE BELIEVES in getting up and working. He says that's the best thing you can do. If you sit around and be lazy as a kid, you don't ever learn how to work. So, for instance, there's only about two of us in our whole grade that do chores after school, and I don't think it hurts to do them, now that I look back. I don't know how to do a lot of things in the house, I'm kind of the outside one in my family. I'm more or less a tomboy.

I don't think women are treated the same as men. For instance, around here, in the group that we travel around with,

CHRISTA LAWRENCE WITH BREEZY

like the Elliotts and Jeff Taylor, I was always the only girl, and so, it was always, "Oh, she's a girl, she can't do that." Or, say, if you're cleaning fries [prairie oysters], it's "Oh no, most girls wouldn't do that." I don't know, it just seems like they think that they're the cowboys and you're just the girl that cooks and cleans. I'm not. They could probably outcook me!

My mom is involved in running the ranch. She's like a hired man, she's always outside. She and Dad work together. My role is more like my dad. Nobody in my family is really interested in showing cattle or that end of it. I used to be in 4-H, in horse – my dad led it – and then I went into crafts for a year, and horse, then I went into horse and beef, and I just took a heifer. Then I kind of got interested in it, and now for a school project, for a credit, I'm actually doing 4-H. So I might go to a packinghouse to learn how to grade cattle. I find that interesting. I've done lots of shows, and actually, this week I'm going to Agribition. I go for the whole week. I'm going with Blaine Brost, from Irvine. I'm working for them for the week. I don't know if I'll be showing, I think Blaine and his wife do most of the showing, but if they're short, I'll help with that.

My parents don't mind what occupation I choose, as long as I find something that I like. I'm going to go to Olds College for the first year, then I'm going to take farrier science. Dad, he's all crippled up and he's a farrier, he says he doesn't know why I want to do it, but I'm kind of interested in that. And then I think after that I'm going to go to Vermilion and take animal health technology, an assistant vet. I'd like to work in that field, just as a part-time job. I'd like to ranch in the future, but where or when I don't know.

I do have 4-H calves, but my parents are nice enough to just feed them for me in the morning. I go to school and when I come home I always have to either go riding, or if there's no riding jobs, I feed my 4-H calves, or something like that. And then, I also play a lot of sports. I have sports practice after school, and since I'm a senior now, it's at night, so some days I don't even come home 'til eight o'clock at night, so I don't do any chores. But then, for instance, my parents will be gone tomorrow, so I have to get up and do the chores before I go to school, then go to school, then come home and do all the chores. On the weekends, I get up not all that early – that depends on if we have a job or not. But my dad usually has a riding job in the summertime, or fall and spring. And I get up and help him, then we either do chores or fix fence or whatever needs to be done. Then in the wintertime I always go out and help feed calves. Skidding logs out of the bush with our team, Hank and Bud, is definitely one of my favourite winter events. We do chores with the team all winter.

I have an older brother and sister and a younger sister. I never worked with my older sister because she's twenty-eight already, she's been gone for ten years. My older brother, I worked with him a lot, and he says he worked a lot more than I ever have or ever will! But my dad, according to them, has got a lot softer. My brother has allergies, so I don't think he'll come back to the ranch. When I was little, from the time I could get outside, I used to help my dad, and I've always been interested in the ranch. I rode at six months with my mom. Later I had a little pony, and then I had another little horse. Then I started barrel racing. And I kind of barrel race all summer, and ride all summer. My little sister, she does the chores that she can do, she can't lift very much yet. She's twelve and she's small. She rides lots and we do a lot of jobs together, like move cows by ourselves. And actually, in the summer my dad'll be out baling and he won't see the cows for about two weeks, so we check them.

It's up to me, and my sister, to take over the ranch – we're the fifth generation, so it's kind of a family tradition. We're both interested in it. I can't really speak for my sister, I'm not positive how interested she is, but I am. I think this is where I want to spend my life. We've thought about tourism here. My sister's kind of the tourism person, so we sometimes get talking and she says, "I can run the tourism and you can run the ranch, and I'll just run around on your land, showing people about." And that's one way it could work because we're not a big enough place to sustain a bunch of families. We only have two hundred, two fifty head of cattle – it depends on the year – and right now, we have even less than that because of the drought. So the way economics are today, even our parents are having a hard time because neither of them works outside of the ranch. And if you did work outside of the ranch, then you'd have to have a hired man, and is it worth it to have a hired man if one parent's driving to town anyway? So, it's kind of a hard place, you've either got to buy more land or one person work off it all the time and modify it so that you could just have one person ranching full-time.

I think a ranch is one of the best places to raise children – that, or a farm, because when you're a little kid, you can run outside and play, your parents don't worry about you. You get into wrecks and your parents don't know about it, you just teach yourself life lessons. And you learn at a really young age how to get up and do chores and how to take care of the cats and dogs. I really liked building things when I was younger. And I had a dog, Jake. He was the best dog ever and I'd lead him everywhere. I had that fence that you can build around flowerbeds and I used to make barnyards with that, and I had this toy cow – I gave it shots and I had an ear tag in it. I'd tie up the dog and pretend it was my horse. When I was only four years old, I broke my own little baby 4-H calf, and I spent a lot of time with it. I liked tobogganing a lot – now we don't seem to get as much snow in the winter, but in the front yard, where the raspberry bushes were, we used to get high enough drifts to make tunnels in the banks. My mom never worried about me. I never went to playschool. Now I play basketball and I go to 4-H functions. I'm the president of our 4-H club and we have a Multiple and a Beef club, so you get together with the horse kids too. And I go to all the brandings and weanings I can with my dad. I just joined the High School Rodeo Association, so I barrel race lots in the summer and go to rodeos.

Actually the distance to town hasn't affected my leisure activities at all because my dad didn't play any sports and he really regrets that now. So he said, "If you want to try any sports, go ahead and try it." So I've tried volleyball, badminton, curling, and I was a long-distance runner in track. So I've really done absolutely every school sport there was. Distance for me wasn't a big deal because we have a really good road, but I know there are kids that can't do it because they're so far away. As for leadership at school, there's kind of popular people that worry about how they look and how they dress, and that's always how it'll be, they do their own thing. I don't fit with them. Actually, I have more boy friends than girl friends. I find boys easier to get along with.

I think ranch women in my grandmother's time stayed in the house more because there were hired men – you could actually afford hired men. And the work was harder, there was more manual labour. Now, it takes a lot bigger place to hire a hired man, so women are going outside and doing the men's work. In a way that's a change for the better, but then, it's not, because the housework gets left undone. Sometimes it's like the mother has to be with the kids, getting ready for school or making the meals and then go outside on top of that.

I think the best part about ranching is you're always with your family, your mom's never here or there, you always eat meals together. Well, my grandma lives across the yard so it was like daycare when we were little. And, if you're having problems with your parents, you can visit your grandma. I think it's a really good place to live. If you want to run away for a while, you can go to your favourite spot, and you're just fine.

Christa Lawrence became Maple Creek Pro Rodeo Miss Cowtown Queen 2003. In November, 2004, she was crowned Miss Rodeo Canada.

PANSY WHITE-BREKHUS ON GOLDIE,
HER HORSE USED IN JUMPING COMPETITIONS, 1946

PANSY WHITE-BREKHUS

I think it was born in us to be ranchers.

Pansy White-Brekhus is the second daughter and middle child born to Walter and Marjorie Humphrey. Although she has spent most of her life ranching in the Cypress Hills, Pansy presently lives with her husband, Sigurd, on an acreage just south of Maple Creek, Saskatchewan. She is widely known for her horsemanship and her exceptional ability to train dogs. Now in her eighties, Pansy is still riding and continues working with animals.

MOM WAS REAL GOOD with animals and Dad liked everybody to chip in and help. We all gathered and handled the livestock. As soon as the baby colts were big enough, we'd halter them, lead them, put blankets on them, and when they got big enough that they could pack us around the corral, we'd ride them. We broke horses all our lives, us kids. I think it was born in us to be ranchers.

When I first started riding I had an old horse of Mom's, and I'd ride round the house in the fenced yard. We even tried driving sheep. We drove dogs and had races on the creek. We would lead the dogs up the creek quite a ways and then start them, and Mom would give them a call and we'd race them back down. That was on the Whitemud Creek. Others call it the Frenchman now, but we don't. My sister was two years older and we would each have a dog. We braided traces and lines out of twine, and Mother would help us make a breast collar out of leather.

My brother was ten years younger and he was born at home. The old lady that lived next to us couldn't talk English – she spoke Norwegian, and she came down. She was a midwife and she was pretty capable. And after, whenever she saw my brother, she would say, "That's my boy."

I went to the Whitemud School, which was three and a half miles away. My sister and I each rode a pony and it was quite a jaunt, especially in the wintertime. Mother got us Welsh ponies and that was my start in horses. Mine was good but my sister's was a little more crabby, but then she was more crabby, so I guess it worked out all right. When my brother started, he had a little Morgan and Welsh pony cross, and she was very nice, too.

Once Dad had some mares he was pasturing for other people. One mare died and she had a colt. The owner said we could either kill it or keep it, so I got the colt and that was my second start in horses.

One day at the rodeo in Maple Creek they wanted my sister and I to do tricks with our ponies, and I think it was Mr. Caton who took up a hat to get a little chicken feed. It didn't matter how much you got, it was money, and the jingle in the pockets. We'd make our ponies lie down and sit up and lay flat and kneel and shake hands.

In the yard at home we staked out a place with little pegs, and we put twine around it and then put ribbons on it. And we had a circus out there. Mom and Dad would watch us stand up on our ponies and ride them or turn backwards on them and ride. We did that quite often. It's a wonder we didn't climb corrals with them, but we didn't. For jumping we had them in a circle and ran them around over rails. As they jump you can bring the rails a little higher, whatever you think they can jump. Then we got going to the horse shows in Maple Creek, jumping and chariot racing.

We had hounds and I hunted coyotes for years. I usually skinned them and stretched the furs out and then sold them. Once we were out coyote hunting and we found this baby deer all alone. We carted it home and raised it, and it would go down in the yard and lay under the currant bushes. He loved to come in the house and he'd walk all around. He just came and went, would go a quarter of a mile west and eat and hide in some wolf willow bushes in Gilchrist's. Then he'd come stalking back at night.

One morning I was going coyote hunting out across the hills, and I went across the top of a coulee and the little mare's feet went out from under her and she went flat. I got back in the saddle but I couldn't use my leg for quite awhile after. I rode home about a mile and Mom bandaged it up and I suffered through it and got over it.

One December I drove a team to Robsart, about ten miles away, for a Christmas concert. We had a full-length wagon box on a sleigh, and coming home I had to turn off on a side road. There was a spring to cross and the team cut too short and I tipped the sleigh over the side of a culvert and upset the family, which didn't please Dad too much. I got the team stopped right away and we got everybody put back. But coming home after a dance, at about three o'clock in the morning, you know, nobody's too bright, but they sure woke up then.

For a number of years I ranched with my folks. I broke away from home and worked out one summer. Then I was down at Beechy for two years, where I cooked and worked inside and outside.

My mother was good to look after my daughter when she was little or if it was too cold or stormy to take her outside. In the spring, my daughter would come out and sit in the pen and play with the baby lambs. Sometimes I'd find her sleeping, using a lamb's head for a pillow.

My great-grandson lived with me for two years. He was my boy from day one. I'd take him to school with the truck to begin with, and then he got so he rode the pony. I sure missed him when he left.

Mother could train dogs most anything, and I guess it just rubs off. My German Shepherd I had from day one and she'd do anything for me – pack a pail, go fetch wood, fetch my gloves from the barn, catch calves for you, anything. She was with me from daylight 'til dark and it didn't matter what you did, she would help you with it. Or if you made her stay back, she would stay where she could watch you. She never got out of sight. This last one I had was a pup off her, and she was the same. When they're gone, it's like having your right arm gone.

Despite the changes, a ranch, I think, is out of a lot of trouble. You don't have so much tempting you – to go jump in a car and rush down the street or to go in a store and help yourself to some candies. The only thing is nowadays there are these push-button things that eat gas. We didn't have that. We had grass-fed motors.

TRIBUTE TO A LADY RANCHER

by Doris Bircham

On the banks of the Frenchman River
  in the heart of the Cypress Hills,
there's the scent of sage and wolf willow
  and the wind blows with a strong will.

It's here Pansy White has been ranching
  since childhood, through each season's change
on the land her father homesteaded
  miles from town on the short-grass range.

When the mower broke down she fixed it,
  she fed cows at forty below,
rescued newborn calves and lambs
  caught out in a late spring snow.

And many of her years spent ranching
  she ran the place all on her own,
put up feed, fenced, shod her own horses,
  never lonely though often alone.

While their parents worked in the hayfield,
  Pansy and Daphne learned to ride.
They practiced on colts, calves and sheep,
  anything with four legs and a hide.

Pansy's special way with her horses
  as she trained, put each through its paces,
won her prizes at shows, in jumping
  and her favourite, the chariot races.

She shows me an album of pictures,
  her Belgian team, each with a blaze,
her dog nursing two pigs and one pup,
  the orphan deer she rescued and raised

the shed roof where Canadas nested,
  where they hatched their goslings each spring,
and she tells me about the pigeon
  who for seventeen years would wing

its way down to land on her shoulder.
  There's a photo of chaps she made,
Welsh ponies, Morgans, her thoroughbred stud,
  horses riding off every page.

The snapshot of her faithful black dog
  who, when Pansy was away, stood guard
and three times got its face filled with quills
  while trying to defend the yard.

She mentions her Holstein milk cow.
  "I hated to milk her," she said.
"She gave so much milk I decided
  to let her nurse four calves instead."

When a heifer calved and then prolapsed
  Pansy said, "She's stayin' right here."
That heifer healed up, made a fine cow,
  dropped a calf on the ground every year.

Pansy, who now has great-grandkids,
  raised a daughter all on her own
and passed to her that same love for critters
  both she and her mother have known.

A sketch of her dad's homestead cabin
hangs on Pansy's kitchen wall,
an R. D. Symons original,
drawn before the logs started to fall.

I have wonderful neighbours, she tells me,
as good as any you'll find,
and they've come to lend me a hand
sometimes when I've been in a bind.

Pansy once had a calf that strayed
to my place, says her neighbour, Bill.
I rode out on my horse, took my rope
and to show you the kind of skill

Pansy has – she came in her truck –
told her German shepherd to go
catch that calf, which it did, by one leg
then we tied the calf ready to load.

"Did you have any wrecks?" I ask her
and she tells me about the team
she was breaking and how they took off.
No matter what she tried it seemed

they ran faster 'til she tugged on one rein,
got that runaway team turned about
and headed them straight for a bog
then spent six hours digging them out.

And one time she was milking her cows
when a fellow showed up at the door,
asked for help, and she told him she'd ride
just as soon as she finished her chores.

So he saddled her big horse to go
  and though Pansy was on the alert,
knew her horse didn't like that man much,
  still she ended up biting the dirt.

Pansy, now in her eightieth year,
  has married a man who shares
her passion for horses and riding,
  and two summers ago this pair

rode with the Boundary Commission.
  Now they've chosen to winter down
below the hills on a chunk of land
  about a mile and a half from town

with their chickens, replacement heifers,
  pregnant donkey, colts and two dogs.
Pansy's hoping to buy a zebra
  but she's sold all her sheep and hogs.

Tomorrow they're off to wean colts
  and in May during Stampede Week,
she'll hitch a pony to her donkey cart
  and drive up and down Maple Creek's streets.

"I just do whatever needs doing,"
  says Pansy, and she's doing it still,
raising Limousins and riding her mare
  in the heart of the Cypress Hills.

MARY GUENTHER

I'm not sure about grieving. I guess it's a hole that never really gets filled but, you know, life goes on. Every time I managed to do something I realized that it wasn't that hard, and it was a little easier next time.

Mary Guenther, born in 1929, is the oldest of Ross and Dorothy Haigh's four children. Mary faces life head-on. She ranches at the top of the Cypress Hills, twenty-two miles south of Maple Creek, Saskatchewan and approximately fifteen miles north of Fort Walsh. She looks out her kitchen window to the north, where the rolling grassland seems to go on forever.

I GREW UP ON a ranch. Our buildings were four miles south of Walsh, Alberta and the place ran quite a few miles south of that. My great-grandfather, Joe Grant, settled there and my mother grew up on that ranch.

They mainly raised sheep but they raised a lot of Clydesdales and Thoroughbreds. They sold remounts to the police and the army and my mother remembered when she was nine years old, sitting up on the corral and watching them buck out horses for the first Calgary Stampede. And I have a program of the first Calgary Stampede in 1912.

MARY GUENTHER AT AGE 15, MOUNTED ON FLORA, 1944

There were twenty-two rooms in our house, not counting closets and hallways. It was actually built by my great-grandfather and he had the house-warming in October of 1903 when my mother was six weeks old. The house had stone walls and then brick, and there were three fireplaces in it. It was built into a hillside so the bottom floor was a kitchen, dining room, laundry room, and then you only went down about two steps into the cellar, so there was a cold cellar, another large cellar, and a meat cellar with thick walls where we hung meat. In those days they had a Chinese cook. On the next floor there was a kitchen and a pantry, a butler's pantry, a big dining room, a living room, a huge front hall, and an office, and that's where the family lived. Then there was a back stair and a front stair. That front stair I dusted many a time and got sent back to dust again, but on the third floor there was a big old bathroom with a tub, and there were five bedrooms, quite large ones, and a couple of back bedrooms – this bathroom that was sort of divided off by a doorway. There was a finished stairway up to the attic – it had big double doors and he had a billiard table up there. Then there was a narrow little stair that went up onto the roof, and we loved to take our visitors up there. You could see all over the countryside.

Anyway there were quite a few people involved in the place. In 1918 they had a bit of a disaster with the flu epidemic, and the two oldest boys, Jack and Harry, died four days apart right there at the ranch, and three hired men. They lost five people there in a matter of about ten days. I guess my grandfather really never got over it.

My mother became a teacher. She taught for about three years, married my father in 1924, and they more or less took over the ranch although my grandparents continued to live there. By about 1939, when I was ten years old, we went out of sheep and went straight into cattle. We had got quite a few cattle anyway and it was no longer easy to get help – a lot of ours had joined the army by then.

It was quite a large ranch and we always had hired men and a hired girl. Sheep require a lot of labour, believe me. We always had shearing crews or something around. Hayed all summer, I'll tell you, but I like to think I lived a different life than a lot of children. I didn't go to school until I was ten and a half, and I think I got a terrific education because the finished attic was full of my grandfather's books, from floor to ceiling. When I could sneak away from my mother – she always had a job for you – you could always carry wood or carry water, but she rarely wanted to walk all the way to the attic. There were three flights of stairs up, so I'd go up there and sit in the window seat and read and read and read. I always say I was practically the only ten-year-old that started school having read *Richard III* and *The Rise and Fall of the Roman Empire*. I'm sure I didn't get out of it what I would today if I went back and tackled it again.

Our family was very spread out. I was fourteen years older than Janie and ten years older than my brother, David, and four years older than Anne. Mother would have loved four boys instead, I'm sure, but it didn't work out that way.

When I was five, my mother was in the hospital in Medicine Hat. She was gone for quite a while and we had two sisters working for us, and I think I drove them absolutely nuts, following them around with books and wanting to know what this said and that said. So Lena taught me my sounds, and by the time my mother came home I was reading to beat the cards. She was a little startled, so next time she was in the Hat she bought me a grade one primer and I read through that in half an hour and gave it back to her, and she got me a grade one reader and I went straight through that, and she got me a grade two book and at that point she gave up.

So when I was eight and a half she discovered correspondence lessons – had no idea where to start me – was sure I must be miles behind, which I wasn't. I did grade three and grade four with correspondence lessons – five and six in Walsh, and then in 1942 they bought a house in Maple Creek, so I went down there but went home to the ranch on weekends.

My grandmother stayed with us at times – she was at the ranch. And quite often Anne and I stayed alone – I'm sure that isn't even considered legal today, but we did. Naturally we didn't throw any big parties or anything. I guess the worst that happened was when we knew Mother was coming down for us on Friday, we'd skirmish around and pick up the dirty clothes and get the beds made. I was thirteen when we went down there and Anne was nine.

Music was a big part of our childhood. Grannie used to get a magazine in the mail called *Etude* and it was very complicated music. I've got it packed away – I can read it but slowly. And I remember when *Etude* came, she went hiking off to the piano and she'd sit there and play right through just like I read *Maclean's*.

My parents left the ranch and bought this place from Gordie Kearns in 1946. Chuck Guenther and I were married in 1948 and we lived at Merryflat for about a year, then we moved down here. Chuck's parents farmed at Merryflat. We had six children, five boys and a girl who thought she was a boy for years.

David was born in 1949, Jack in 1950, Harold in 1952, Glen in 1954, Mary Jane in 1956, and Richard in 1958. I was an absolute walking population explosion. And I wasn't even Catholic. But anyway Chuck was killed in a tractor accident in 1959 and the decision was – would we stay or wouldn't we, and I'm glad we stayed. I think it was a better place to bring kids up.

At least here when I went out and raked in the hayfield I took the kids along and built a bale house and they played in there and, you know, I'd rather do that than live in a city where I had to leave them at day care. They sort of grew up around me and they were all riding by the time they were about three.

I'm not sure about grieving. I guess it's a hole that never really gets filled, but, you know, life goes on. Every time I managed to do something I realized that it wasn't that hard, and it was a little easier next time. I think my children missed a lot by not having a father, but on the other hand, everybody had a horse. They used to take a sandwich in their pockets and head off about eight o'clock on Saturday morning and they might go to Fort Walsh. They might ride down to their cousins at Merryflat, which was fifteen miles away, and they might get home at eight or nine o'clock at night and I didn't worry about them too much.

My father was still somewhat involved here as far as business went, and we had bad roads in those days in the wintertime. My mother was very good to keep the kids when they had to stay in town after school. And Lloyd Coleman lived down the road, and he and Irene babysat many a time for me. Lloyd was an older man, and he wasn't exactly a mechanic but he could always make something run. So I can't say I did it alone, but I had the main responsibility. And we had a lot of hired men. Some of them I remember very fondly and some I don't. My father felt like, "Now hired men aren't going to like to work for a woman," and some of them don't. And I was kind of glad when my kids got old enough to do quite a bit.

We broke a few bones, the odd collarbone. With six children as wild as March hares, each would fall out of the hayloft or get cut and need stitches or get bucked off every once in awhile but on the whole they survived quite well. Mary Jane wore glasses and we weren't exactly rich. She had a little horse that bucked every time she got on it and then it would be away for the day, but everybody would yell, "Mary Jane, where's your glasses?" They were always flying off somewhere. Somebody said to me one time, "I suppose the boys spoiled their only sister rotten." I said, "Sure, every time they wanted to snub a horse up they were breaking, Mary Jane was the one that got on it."

There were no double standards in this family. If I wasn't going to make the boys do dishes, I didn't make her do dishes. If the boys were going riding, she went riding. I felt right from the start my daughter was going to have to look after herself just as well as any of the boys, and there wasn't very much her brothers ever did that she couldn't. And when Mary Jane tells her brothers something, they listen, every one of them.

The first time I ever went to an RM meeting, I was the only woman there. I had to take a deep breath because I didn't feel like pushing in and men sort of moved away a little bit and it was a little lonely on the start but that doesn't happen anymore. I see young couples today and they're pretty well partners and it's the way it ought to be.

When you think about it, women just got the vote about ten years before I was born and that was just given to wives and mothers and service women, to start with, I believe. I think of Emily Murphy, in 1928, finally declared a person by the British courts. That was the year before I was born.

I can remember a Stock Growers Convention they had in Maple Creek years ago and my mother loved to entertain and did it beautifully, so she had a ladies' luncheon for those wives and I think there was a bit of a fashion show that afternoon. I went in and helped her, but I remember thinking, "This is ridiculous. Why would you bother going if you can't go to the meetings?"

In the sixties I got quite involved politically. At that time we were in the Shaunavon constituency. I was membership chairman, and then they changed the boundaries and I was in the Maple Creek one and I was president of that for a while, probably not a terrifically good one, but anyway ... I think I set out to save the world in my thirties, but I kind of gave that up. The first political meetings I went to I was probably not the only woman, but there weren't a whole lot of us.

I've put in a lot of time and a lot of miles – I've been on the Sasktel Board of Directors and I enjoyed doing that. I met a lot of nice people. I've been president twice of the Old Timers Association. I don't do a whole lot about that anymore either. I was on the executive for years, but I go in every once in awhile on Sunday and babysit for the day, and I have been scanning a lot of the pictures and trying to do a few write-ups to put in there because we're hoping to have a bit of an archive. Fifty years from now somebody can come and perhaps find family.

I really needed to get going on the computer with this family tree program, so I started this spring and thought, I'll put in a couple of days and get it done. I kept thinking – who else is going to do it if something happens to me? I'm up to darn near twenty-two hundred names. It took me weeks. The earliest one I've found in my family is from 1509.

I have been very pro-choice for years and years. I used to write letters to *Chatelaine* and I embarrassed my mother and my mother-in-law considerably, but I felt it was a very important issue. I knew a girl and during the summer holidays she had an illegal abortion and died. And she was a very nice girl, very quiet, a little shy. I don't think she had the kind of family that would have accepted it in any way, and she must have been desperate. I would prefer prevention, obviously, but I have always thought it should be available and nobody else has the right to make the choice for you.

And I'm all for breastfeeding. I got talked out of it with Davie and I've always regretted that. Davie weighed ten and a half pounds, was twenty-five inches long and skinny as a rail. Everybody assured me there was no way on earth I was going to feed that big child. My mother thought it was really silly. I got no encouragement whatsoever around the hospital – didn't know what the hell I was doing. Davie had colic and nothing seemed to agree with him, and in those days they had every kid on, I think it was Carnation, and the next year when Jack was born it was Farmer's Wife. By then I'd learned a lesson or two, so I nursed the rest, most of them for a year.

I'm healthy as a horse, actually until, well, nine years ago I had a slight bout with breast cancer but they caught it very early and I've never had a sign since. I just had radiation, not chemo. I've had two cataracts removed, probably from being out in the sunshine a lot. My mother had cataracts but not until she was in her eighties – here I was, by the time I was seventy.

I like being outside. I've done a little bit of everything in my time, I think. I was never a great mechanic, I'll tell you – if somebody showed me how to do something – step one, two, three, I can do it. That's something I missed when Chuck was killed – my God he could take something apart and put it back together.

Now I go out to look at the cows, help Davie in the corral, and I go out and help fence, but I'm not as busy as I used to be. I can remember we put up prairie wool all summer, and every fall we stacked about ten thousand small bales – I've stacked a hell of a lot of those, I can tell you. I still go out with Davie when we feed in the winter. Now we have the big round bales and just haul them out, so it's a little different story than it used to be.

You know, I think I've probably been giving the impression that ranch women are much more equal and involved and so on today, but I've been thinking back and I think women on farms and ranches have always done what needed to be done. There are an awful lot of those women that didn't just cook for the threshing crews but worked in the field and milked ten cows and drove teams and whatever, so, you know, it's just that it is more official today or recognized or acknowledged possibly. I don't think people have changed that much. Circumstances change somewhat and ways you do things but I think there's always been very strong women.

ANN SAVILLE

We always milked cows and it was so nice to take a can of cream to town and be able to have enough to buy groceries.

Ann and Bill Saville, both born in 1923, retired from their ranch and moved to Eastend, Saskatchewan in December, 1999. There is much coming and going at their home, with Bill operating his harness shop out of the basement and Ann involved with her many community activities upstairs. Here one finds western hospitality at its finest.

I WAS BILL AND ANNA CATON'S youngest child, born on my mother's birthday. I grew up on the VL Ranch on Farwell Creek in the Cypress Hills of Saskatchewan with a half sister, a sister, and a brother. It was isolated like any of the homes were in those days, but it was good.

Dad had batched so he was a good cook. He got up and he always made his quite renowned sourdough pancakes. And Mom always made chokecherry syrup. We had chickens and there were always men to feed, and we used to eat the traditional pancakes and eggs and syrup and gravy or whatever. Horrible breakfast.

ANN SAVILLE SPINNING, ABOUT 1970

Our place was twelve miles from Robsart and forty-five miles from Maple Creek. We used to count – twenty gates, I think – there were no cattle guards. We hated going across country on that Davis Creek Road. It was quite an undertaking in the old Model A. We didn't often go to Maple Creek – we didn't go anywhere, you know. Mom and Dad stocked up on flour and sugar and they canned jars and jars of meat, vegetables and fruit.

There were no schools. I can remember Dad talking one day to Mom. He said, "Isn't it funny? We pay seven school taxes but there's no school." He couldn't understand that. So, we used to get government correspondence.

But, first of all, when I was in grade one, they had rented the old hospital building in Robsart, a big old three-storey building, and we lived there and Mom took in boarders and we had the whole upstairs for our toys and playground. It had a big skylight because it had been the operating room.

The next year was '29, the year of the crash, and we moved back and started correspondence. Mom would teach us and we'd have school from 9:00 to 12:00. It was perfect. We had the afternoon to ourselves – we could go outside. Finally, when they figured I could ride – it was three or four miles to Clear Valley School, which was west of us – then we all started riding to school. Grade nine I took at home.

In the drought years or "hungry thirties," Dad got a job managing a pasture at Carberry, Manitoba and he moved his family and cows there. I took grade ten there, and grade eleven I boarded in Maple Creek. The next year Mom had to go to Medicine Hat for surgery. I stayed home, borrowed some books, and with correspondence did half a year of grade twelve. I did the other four grade twelve subjects at Robsart the following year. The war had started when I was in Maple Creek and at least four from our grade never came back.

Then I was home for a year or so helping Dad because my brother, Jack, had gone overseas. Dad went to a Stock Growers meeting in Saskatoon and got talking to Grant McEwan, who wanted to know what I was doing. Dad had told him I was just at home. And actually classes had started but he said if I got hustling up there I could probably get into first-year agriculture at the University of Saskatchewan, and he did get me in. At that time there was one girl in each of the first, second, third, and fourth years. I always used to think how lucky I was to be there.

In the final year I got the Tommy Frazer Memorial Trophy. It was the first time a girl had won that, so it was a great honour. It was supposed to be academic, sports, and just general spirit, I think. I represented three colleges: agriculture, law, and engineering. Later I was the elected representative for Maple Creek and Shaunavon university grads on the University Senate for two three-year sessions.

After graduating, I worked a year at the Dominion Experimental Farm at Lethbridge, where I met Doreen, the artist. There were eleven in her family and her youngest brother was coming to the rodeo, and that's when I met Bill. He rode saddle bronc at the rodeo. Doreen and I were riding in a musical ride so were at the grounds both days. Bill, the youngest in his family, grew up on the Z-X Ranch east of Wainwright Buffalo Park in Alberta. And Dad always warned me, "Well, whoever you do go out with – don't go with a cowboy." So here I was, and I saw him again on July 1st weekend at Swift Current, Saskatchewan, where he won first money in the saddle bronc event. But it turned out all right, anyway. Bill and Dad got along just fine.

We got married in 1950. Dad had added the Bacon place to the VL Ranch in 1947 and offered that to us as part of the Caton Cattle Company. After our honeymoon, we arrived home at the Bacon place and were met by an old, old gentleman brandishing a twelve-inch butcher knife and clad only in his long winter underwear. He introduced himself as Mel Elsom and explained that he had homesteaded with the

Bacons' and had arrived with a threshing outfit from the United States. He had no home to go to, and could he live with us? We both realized this could be a real bonus when he explained the water system. Water was piped from a spring about thirty feet up the coulee and flowed through the house, with the overflow going on down the slope to the barn and corrals. The heating system consisted of a large steam boiler encased in brick and built in the dirt cellar. It had the potential to blow up or to go out and freeze everyone plus the water system. He also had some tricks up his sleeve on how to get heat from the soft lignite coal, found just a mile north of the house, and he loved to cut up wood to burn with it. Often at any hour of the day or night we could hear the old bucksaw and Mel singing up a fine tune together.

On another occasion Mel had gone to bed one night, carrying the usual coal oil lamp. His room was just off the kitchen and suddenly there was a funny gurgling, choking noise and a bang. No one had mentioned that Mel was epileptic, but we knew there was some sort of problem. Bill grabbed a gas lamp and I got a jug to get some water from the bathroom, when I realized Bill was flat out on the floor in a faint. The lamp was still upright on his chest. Why we didn't all burn up, I don't know. There were no serious effects and Mel was the least perturbed, but I never did let Bill forget, as a joke, what a hero he was that night.

The house was a lovely old log building originally from the Farwell Creek NWMP outpost situated in the valley near my old home. The Bacons had purchased the military buildings on site, numbered the individual logs, hauled them to the Bacon place, and rebuilt the buildings. The main living quarters were 24 x 36 feet and the top central beam was a full 36 feet in length, with a circumference of 36 inches.

The only way we got around to start with was with horses. The teacher always boarded with us so we never really were alone, as there was always hired help, too. We kept busy and we really enjoyed what we were doing. We had each other and, in six years, what we considered, the perfect family – David in 1951, Susan in 1953, Jim in 1955, and Nick in 1956.

I can remember Sam Wright coming out with Dad one day, and when he learned we had only thirty-five calves to sell that fall, Sam just shook his head. He said, "They'll never make it," because that wasn't enough, but we increased them and bought more land and of course, Bill always kept horses and we made money with our heavy horses.

We always milked cows and it was so nice to take a can of cream to town and be able to have enough to buy your groceries. Anyway, I was getting so sick of this ruddy separator, you know, and we were in Havre at the International Harvester shop where Bill always did his business and I saw one of these automatic washing machines, separators that wash themselves. Finally one day I said to Bill, "I think I'll write to that IHC in Havre and see if they still have that." I can't remember how much money it was but we sure didn't have much money to buy it. They got right back to me and said it was for sale now, second-hand. We got it half-price.

I helped with the summer fallow – it's such a dirty job. Haying was good. I mostly just raked or used a sweep where you have a horse at each end of this sweep and then you push the bunch ahead, the bunched windrows. I remember one day I was raking and the poor old seat must have weakened, and suddenly my seat was gone and luckily the horses did stop for me. Bill was up on the stack at that time. He saw me go over top of the rake teeth. That was kind of scary.

I went off a horse once – a year after we were married. Bill always broke horses and he'd gone over to Jack's, to my brother's, to pick up a colt that Bill was to break. We'd ridden our two regular horses over and I was leading his saddle horse home and he rode the colt home. We were down in the bush and something spooked his colt and it started bucking. Of course, the one I was leading got scared – it wasn't a real gentle horse, either, and it pulled back and somehow I must have jerked my horse. I don't know yet what happened. Anyway, I went over backwards and I guess I landed on a rock pile and I was out for a while, but Bill said I got up and said I was okay and I don't remember any of it. But he talked to me and we just walked and led the horses, and I told him I could ride so we rode home. It was only about another mile home through the bush and I can remember just as we rode up to the barn – all of a sudden my mind came back. I said, "When did you start riding Danny?" That was the name of this colt that we'd picked up at Jack's and I didn't know where this horse had come from. I didn't realize but found out later it was my back – I had broken little sharp pieces off the vertebrae, so I went through a bit of misery. The doctor planned to put a graft from my leg onto my spine, but it was too broken. There was nothing to graft to so he just cleaned it out. They thought I would be on the striker frame but I wasn't, and I was only in Regina for two or three weeks before I was home. Then I had an umbilical hernia and a couple more hernias. I don't know why I fall apart but I do. Then I had to have a hysterectomy. I guess I'm a wreck, but I'm okay.

I was still riding right up to near the time David was born and I don't think it mattered a darn. He was sure a skinny little baby – Mom and Dad didn't think he'd survive. He only weighed five pounds, two. He had no hair – he wasn't a very healthy-looking baby but maybe that's because I didn't give him a chance to grow. I don't know. Anyway, all four survived. I guess we were lucky. They were pretty well on their own. People would come along and say, "You know, your kids are away down the road somewhere," and I would say, "Well, they're okay." And they were and they said they had a wonderful life to remember, just out and about and riding horseback and playing in the creek and all that.

Our children started at Farwell Creek School, which was just half a mile down the creek from us. When Nick, the youngest, was in grade one or two, they were bussed to Ravenscrag and later to Eastend. Nick was always carsick. So we tried gravol and he said it didn't work because by the time he really woke up from gravol, it was time to get on the bus to come home again.

I had a phone call from Regina when the kids were older and was asked if I would sit on the Farm Debt Review Board, so I did that for ten years. It was a real eye-opener. I didn't realize how serious the situation was for so many farmers, and of course, we were all over the province, so it was very interesting and educational.

After the Farm Debt Review Board, the provincial government set up the Farm Land Security Board and they asked some of us to go on it – they did exactly the same kind of work, mediating between farmers and financial institutions. Quite a few people we worked with really appreciated what we were able to do because a lot of them had the idea that nobody gave a damn whether they lived or survived or didn't survive, and some of them didn't. I remember being so shocked by the big hog farms all over Saskatchewan that had been promoted – the government had set them up, and they just went flat broke, so that was sad.

I wrote a few articles for the *Family Herald*, the *Farm and Ranch Review*, and on work being done at the Experimental Farm. It must be thirty years I've been writing up the Ravenscrag news. In 1995 I received recognition for two three-year terms on the Saskatchewan Crime Stoppers Board. I used to do census and those agricultural surveys, too, which the farmers hated me for, but that was kind of fun.

We showed the Belgian horses for quite a few years. The children helped feed and clean out the barn, but once one of them said, "Wouldn't it be fun if we could go on a holiday without the horses?" Because we always said, well, that was our summer holiday, and they did have a ball, especially at Calgary. The years when we went, we could live in the barn with our horses and it was so much fun. We had a camp stove and we did all our cooking and everything, and the four kids and I lived in the box stall. And we didn't worry about the kids. They could wander around the fairgrounds and watch all the exhibits and games. Now you couldn't let a little girl of eight years just wander around. The elephants were in the parade that year, and Jim got to ride on an elephant – the other kids were quite envious.

We should have moved to town a lot sooner than we did for Dave's sake. We had separate houses but it's still not the same – it's very difficult to live in the same yard. It's hard to leave, but I had been fairly sick with this stupid arthritis, not that you really get sick, but you get helpless. And so it was a good time to leave, and then we had the good fortune to win a house in Saskatoon. So we sold it and it wasn't a stress on the place at all. We could go out and buy one on our own, and we did that, so that was perfect. And I never dreamed we'd ever sleep in until eight o'clock in the morning. Six o'clock you had to be on the go. Bill milked cows by five-thirty.

Bill and I worked with 4-H clubs for a while. We had a bed & breakfast and we were always busy with the horse shows. Then Bill and I started square dancing, which was really a wonderful experience for us. We did it for eleven years. We travelled around to quite a few places. Then Mom came to live with us. She was eighty-seven and she was with us for ten years. The last year and a half of her life we had to take her into Maple Creek, as I couldn't manage any longer.

I can't garden much anymore. I don't seem to have any strength in my hands, but my legs and feet are bad, too. I try to keep knitting and spinning a little bit. I used to spin a lot. The Hutterites showed me how and I bought a wheel from them, and then in later years I got a New Zealand wheel, which is much easier to operate because it's a bigger wheel. Mom used to wash wool and we would card it for quilts, but she never spun it.

Mrs. Aslin showed me how to wash it properly for spinning, knitting, and carding. I did a lot of sweaters and mittens for all the kids. They're so much warmer.

At Ravenscrag we had a community club, and Bill and I were both active with it. Over the years I've been awarded a number of trophies, one of the most meaningful being the Spirit of Ravenscrag Award, where mention was made of the value of keeping up the decreasing rural communities. We did two history books, which takes a lot of time, and I was editor of both but I had wonderful help. It was fun. I forget who I asked first if they would be interested, and right away, the whole community was all for it.

And Point View Cemetery – we're still involved with the upkeep of it. It's north of Ravenscrag and all my family are buried there, and since so many have been cremated now – we're still all in one plot – they're just under Mom's big rock. All she wanted was a big field rock, and Jack finally found the right one and had it moved, and there are different plaques that are on this rock.

As for ranches fifty or a hundred years from now, it scares me because everybody knows the little guy is getting squeezed out. It looks like the Hutterites and the Indians and a few big outfits – like we have one who has bought land north, south, east, and west. And maybe it's going to be family setups, I don't know. It just seems to be getting tougher all the time.

In January, 2004, Ann Saville was awarded an Honorary Life Membership in the Saskatchewan Agriculture Graduates' Association of the University of Saskatchewan.

MARY JANE SAVILLE

I love horses and I always liked being the first one on or being the one that started a colt.

Mary Jane Saville is adventurous. She is a good listener, she is warm, and she is decisive. She grew up with five brothers and has worked with hired men and neighbours in and out of the corral. She doesn't expect special treatment because she's a woman. She says, "If I was out there to do a job, I expected to do the job. I have as many men friends as I have women friends. I've always felt safe around ranch and farm men." [Her mother is Mary Guenther and her mother-in-law, Ann Saville.]

I GUESS YOU NEVER appreciate your parents until you grow up. As I look back on Mom's life after my dad died in 1959, she was left with six kids and a ranch. I was three, my youngest brother was nine months and my oldest brother was ten or eleven. Mom decided probably right away that she wasn't going to leave the ranch – this is where she wanted to be and she never gave up. She never said, "I can't do this any longer."

Mom never made me stay in and do dishes. If it was haying time, I got up and went out to the hayfield with everybody else

MARY JANE SAVILLE RIDING A BRONC
AT MAPLE CREEK'S RANCH RODEO

and when I think back – wasn't that an awful thing for me to do? I did do some housework but no more than the boys. She never, ever differentiated between myself and the boys. And I don't know what she could have done about it, anyway.

One instance when I realized I was a girl, I was probably eight or nine. It was in the fall and they were preg testing. We'd spent the morning gathering cows and I got up from the table after dinner, was headed out the door, and my grandfather, who was quite a character and who didn't need to bark at you more than once, said, "Where do you think you're going?" I said, "Corral." And he said, "Not today, you're not." I was so mad and Mom says, "Well, some men are like that." I didn't know why the boys should know whether the cows were pregnant and not me.

I was probably the one that broke more colts than the boys and I don't know whether that was smart or not. I was always the one that they could sucker into it. I love horses and I always liked being the first one on or being the one that started a colt.

I've broken a couple of arms. The last one was off a cow, so that doesn't really count. Once I was breaking a black mare and my brother, David, was ponying me, you know –, had a shank on her so she wouldn't buck away from him. We did that all the time. I guess it's fairly dangerous, I understand now. We were coming home and it was pitch black and he smoked. He got off and opened the gate, and just after he got back on, he lit a cigarette and this scared my black mare. And she took off – at that point he had already given me my shank back and I was just riding her by myself because she was supposedly broke by then – and she bucked me off and knocked me out. It was so dark they couldn't find me, and they couldn't see my black horse. She ran home. Finally they found me – I was unconscious – and it was just after somebody had broken their neck in a car accident and nobody wanted to move me so they phoned the ambulance. I kind of came to sometime in the night and realized I was in the hospital and they wouldn't let me out. Everybody had gone home and left me there and I woke up in the morning, phoned Mom, and I said, "Will someone come get me?" She says, "Oh, where are you?" She had been away from home and come home after everybody'd gone to bed and nobody had left a note saying that I was in the hospital.

I didn't enjoy school. Nothing was hard and I loved to read. I still love to read, but it always seemed they spent way too much time doing things I never was going to be doing.

One winter I worked at Joe Saville's for a couple of months. He had his barn full of weanling colts, so he asked me to come down and halter break colts, but I ended up feeding cows a lot. I had never chopped water holes in my life and down there, that's all there is. You spend about four hours chopping water holes. I suppose he feeds between five or six hundred cows, and you have to water them all on that creek.

I worked in Peace River country for Alberta Forestry for six months cooking. Then I decided I wanted to go into a fire tower, so I went to a fly-in fire tower about a hundred miles north and west of Peace River for five months, and I read a thousand books.

Next spring a friend and I decided to go Britain. We bought bicycles in northern Scotland, cycled down through Scotland, northern England, over to the continent and then to southern France. You stay in hostels and I really enjoyed that, but I decided I didn't like holidays – two months holidays and what do you do every day? So I thought I'd get a job. I looked around France and the only thing I could find was picking grapes. I had picked fruit in the Okanagan one summer and I decided I didn't want to do that again. I went back to Britain and said to myself, "If I don't have a job within a week, I'll go home." I looked in a magazine called *Horse and Hounds* and there were several jobs for stable people, so the first one I phoned – the lady was in London for the weekend with her husband and they said they would interview me. So I went to their club, which was kind of fancy, and they interviewed me and sort of just wanted to know if I had any horse experience and I said, "Well, yes." But that's hard to prove when you're in the middle of London, so they asked me out to their farm for the weekend and I went out on the train, out to Upton, which is out in Worcestershire, which is in the Midlands right next to the Welsh border.

Joyce knew horses. She was a level three instructor. I didn't get wages but I worked for lessons for a year. Rode all the time. I had never ridden English before, and now I actually prefer it. Joyce was married to a fellow named James, who had a fairly light stroke about three weeks after I got there. He wasn't going to be able to ride for the winter, and they had a subscription to a hunt club, which is what country gentlemen had. So, instead of giving up his subscription, which I understand was quite an expensive item, they gave it to me for the year. So I hunted twice a week with Joyce.

She wouldn't let me go hunting for about a month, not until I was able to jump four feet. She said, "The horses can do it, I just want to make sure you can stay on." I thought, they can't throw anything at me that I haven't seen. I'd been riding since I was pretty small – but they can ride and they go through some rough country. The hunt we were with was a very fast hunt – there's different degrees. It's all sorts of farmers and pretty much professional hunt people. The first day we'd been hunting for about two hours and we came to a jump and it was a five-plank fence, probably four feet high at least, and it was running down a side hill. By the time we got there it was really churned up and muddy and there were loose horses and people had fallen off and I rode up to this scene and I said to Joyce, "I can't do this." And she said, "Oh, sure you can." She said, "Well, Jason can do it." That was the horse, a lovely sort of an Irish Cobb that I was riding. And I said, "Well, maybe Jason could by himself but I don't know if I can do it."

She says, "Well, okay, we'll find a way around." So we had to go and look for a gate. We probably rode about two miles and we lost the hounds, so the hunt was done for us for the day and she was mad. She said, "If you ever do that again I'm never bringing you hunting again." So I decided if I was going I would have to go with the flow, and it turned out to be a tremendous lot of fun.

I had been away from the ranch off and on, but I always came home for calving because I loved calving. After I married David, when I first came here we dehorned everything – we were using horned bulls and almost pure Limousin cows, and they'd also been using Brahma bulls on their heifers. They probably had ten half-Brahma, half-Limousin cows at that time. Those old Brahmas were very protective and Bill, Dave's Dad, would rope this old Brahma cow and we'd tie her to a tree and dehorn her calf. But he would always get off and ear mark and dehorn the calf. About the third year I was here, he held the cow and I was the one that got off to do the calf. I said, "Well, it doesn't have horns." He said, "Oh, no, they never do." I said, "Well why do we do this? I don't understand. Explain to me why we just about kill ourselves so that we can put an ear mark in this calf?"

Anyway, Bill loved things wild and I think we got along really well because I liked riding colts. I brought two of mine from home, a four-year-old and a five-year-old. He didn't have any kids that liked to break colts, and we had a lot of fun. I got these two started. One was a mean little mare – I rode her for two years and I don't think there was a day she didn't buck me off. We'd be riding along just as nice as can be and then she'd just blow up and you never knew, so finally we canned her. We started colts for other people and raised some of our own for a few years – always had three or four to ride.

The Belgians started out being Bill's and in the last ten years Dave and I have bought them. Bill was quite a teamster, so we were always going places we probably shouldn't, catching corners on racks and having runaways. You always have one horse you wish you'd left at home, but you can't when you need four to drag the sleigh around. One time we had one horse that was just driving us crazy, so we decided we'd make him really work – we hitched him by himself in the lead and the two behind, but he couldn't really do it. He wasn't as good as he thought he was.

Feeding the cows and horses in winter takes probably two or three hours, and we use a feed wagon with a grapple fork on it and feed loose hay. This is the first year in the twenty years I've been here that we haven't fed with the four-horse hitch. We were kind of waiting for snow and then I had a lame mare, so we were waiting for her to get better.

Animals come first. Your animals basically come before your family even. I've always been taught – they depend on us for absolutely everything that happens to them, for their water, feed, for their medical attention, everything. They're like children, even more so than children. Children can tell us what they need where the animals can't, and not only do we make our living from them but we have a reputation from them as people, how responsible we are to our animals. It has just been ingrained in me since … for generations.

I'm lucky with Dave. I don't know whether he expected me to do as much, but he's certainly let me do as much as I've wanted to do and I'm considered equal. And Dave's parents were always good that way too. I had never been around any farm work before. We didn't cultivate any land at home, so I did the calving and Bill and Dave put in the crop and I helped during combining. There's never been any way of making me feel that I shouldn't do something or that I wasn't equal to doing something.

We adopted Clarence when he was four and a half, and I always say he knew a way more about being a kid than we did about being parents. And Clarence sure changed our life and for the good. He came from northern Saskatchewan, never had seen a cow or a horse before he came. He is treaty Chipeweyan, Dene, what the Crees here call "bush kids." The very first day he was here, Ann, Dave's mother, was gone and we were going to ship our dry cows, so I just put him on an old gelding that we had here and away we went – put a lead shank on him and he came with us.

He's in grade eight this year and I did home-school him for two years. He's a very sociable kid but I think Clarence has an identity problem lots of times because he certainly lives in the middle of the white world knowing he's not white. The Nekaneet Band here have been very good to him. He actually has a Cree name that they gave him. We've taken him to sun dances, and Dwayne Francis is Clarence's godfather, so he's been very good to Clarence and makes sure he gets to cultural things.

Soon after we bought the ranch we bought life memberships in the Saskatchewan Stock Growers because we started to realize how important the work they do is to the ranch community. About five years ago, Dave became a zone director, and one meeting he had something else on and said, "Well, why don't you go and just take notes." And he never went back. I started to get involved, became a director, and then became chair. And the deal was we were going to rotate the chair and I wasn't going to have to take all the meetings in Regina, but I just got so interested in all the work that the Stock Growers do.

We just had our annual zone meeting, and also I did some CCA [Canadian Cattlemen's] work right after that, and it seemed for about two weeks everything was scheduled around that sort of thing. I don't say our outside work suffers, but it certainly gets abbreviated. You rush and do things and just hope nothing breaks down and no disasters happen, and of course, my housework kind of gets pushed onto a back burner a lot of times.

And then two years ago Joan Perrin from BIC [Beef Information Centre] asked if I would be interested in being an alternate on the BIC board of directors. Saskatchewan has two seats and then you have to have an alternate. Then last fall, Wilf Campbell, who had been the chair of the BIC committee, stepped down and I was asked to step up to that. And we discussed that here – how much more that would be. It is supposed to be just three or four meetings a year, but after that they decided since I was going to the BIC meetings, I might as well take the CCA director's position, too. Now I'm sort of starting to realize I may be getting in over my head, but I'll see. I need time for myself, too, so I make sure I have my reading time and my walk time and things like that. I could retire from ranching, but I don't think I'll ever leave the cattle industry behind. I find it to be fascinating even beyond the farm gate. And it's good we have an industry that's directed by people from the grassroots level.

I can't see myself ever living in a city but I could do other things. If I just had horses, I would be okay. One of the things I enjoy is just going some place else and riding. A few years ago, Cindy Zeigler, my neighbour, and I got about ten women together and it was a women's only camping trip, with our horses, to the west block of the Cypress Hills. And she still says it was the best holiday she ever had. We spent two or three days up there and just rode and sat in the creek and drank beer. And each year, I take a riding clinic with Robin Hahn – Robin's been coming to Maple Creek for fifteen years.

I watched the very first time they had a ranch rodeo in Maple Creek, and I thought, "Well, you know, I could do this," and I got talking to Cindy Zeigler and Connie Delorme and Sandra Hanson, who's from Eastend, and Sherry Wright, who is married to Bill Caton. We practiced once a week for about three weeks at a little arena down in Ravenscrag, and we really had a lot of fun. The only way anybody would go in it was if they didn't have to ride the bronc. I said, "That's okay, I'll do that." I mean, I wasn't forty yet, so I decided I could still do this. I'd always wanted to ride a bronc out of a bucking chute, so I was pretty excited about it. The other guys were mostly card-carrying bronc riders. I mean, we were up against Ross Kreutzer. Anyway, Corbett Falkner and I tied for first, each with a seventy-four.

Dave still tells Corbett, "Well, I remember you used to bronc ride and you rode like an old woman." But I enjoyed every second of it. I can remember sitting in the bronc chute and I thought, "Well, I'm going to just absorb everything about this, because I'm never going to do this again." And these men were worried and fussing around me, putting my feet in this way and that way in the stirrup and making sure I had my shank right – and all I can remember – Dave's Dad was up behind the chutes and he used to ride broncs when he was young and he had helped saddle my horse. Anyway, he said to me, "Well, just ride like you do at home." And I was happy the horse was one of the mares from Erna Peter's and Helen Gilchrist's bronc string. I think that year, as a women's team, we really brought the crowd.

Cindy Zeigler and I worked together to put on the "Women in Agriculture" conferences in Maple Creek. It was in the eighties when there was a lot of farm stress and debt. And when I say to Cindy, "I've been thinking," she always says,

"Oh, no, not again." 'cause I always get her involved in something wild, but anyway we got talking, you know, it would be fun to do something where women had a chance to get up and say what they wanted to say, and we wanted them to realize that they're not alone in this. We started contacting speakers and phoned around and got a little bit of money. We didn't want to get government money. We wanted to have this paid for, so we did it as cheaply as possible and I think it was twenty dollars for the day and that included a meal. We would get it all lined out in February so by June when we were busy it would all kind of start working.

I started out going to the Healing Lodge as a volunteer to visit women and to participate as community in some of the events they did there. Then they asked me to sit on the advisory committee – it's a sort of a liaison between Corrections Canada, the staff, the residents, and the community. I feel their direction has changed in the last couple of years and I'm considering stepping down from the advisory committee and going back to just being a volunteer where I enjoyed visiting the women. I think the long-term residents need a sort of stability, someone who comes just for them, because a lot of them are a long ways from home. Some get one visit a year with family.

I always feel very attached to history. I don't think that we can detach ourselves from the last generations, and I think there have been men and women involved in agriculture that are great role models. And there have been many great role models in this area.

As a director of the Saskatchewan Stock Growers Association, I will be attending a Canadian Cattlemen's Association meeting in Calgary this week. There is a lot of frustration in our industry, and when I sit at these meetings I often feel like a minority. People from our province tell me. "But you have to be there so there is some moderation." I know it is true and I know it's important for the voice of our provincial ranchers to be heard.

The SSGA does not receive check-off dollars but is funded through memberships, so when ranchers join the organization, they know the association lobbies for free trade, free markets, and free enterprise. Even when we are downright discouraged and we think no one is listening, we must keep our eyes on those goals and strive to keep ranching a viable industry.

When this wreck is over, hopefully we can wean ourselves from the government handout mentality and we will have learned something so we won't have to repeat this in the future.

Being on the national association has opened my eyes to what a wide range of producers there are out there. I know I have the privilege of representing the most honest, the most efficient, and the most independent producers in Canada, perhaps the world.

ERIN BIRCHAM

I think people are surprised now, if I say I'm from town. They don't see me as a city girl.

Erin (Butterfield) Bircham ranches south of Piapot, Saskatchewan, along Bear Creek with her husband Wayne, and sons, Leigh and Jordan. She was born in 1963, eighty miles away, in Swift Current, Saskatchewan. Erin trained as an X-ray technician. She was a city girl, nevertheless, she has readily adapted to the ranching lifestyle.

I DIDN'T GROW UP on a ranch, and city people in general, when they're driving down the highway and see a cow – it doesn't matter whether it's a bull or a cow or a heifer, or what colour it is, they're all just cows. So I had to learn all these different animals, different kinds, so that was interesting in the beginning. Tractors were all tractors and you didn't know any of the implements and it was just all very new.

I was an X-ray tech, working in Medicine Hat, Alberta. I met my husband, came out here, and I liked it. When I first came to the ranch, I didn't do much, but it was good, it was nice being outside. Now my husband just expects me to do whatever, everything. There's not very much that I don't do. I

ERIN BIRCHAM PROCESSING CATTLE, 2003

don't run the baler and don't know how to run the chainsaw. Well, I used to say that I don't milk cows, but I can't even claim that anymore! There's just nothing that he doesn't think I can do, although I get into trouble a lot. He might be mad at me after the fact, but he always expects me to do it.

When we first got married, seventeen years ago, we used to go out more socially. There's a remarkable difference in my social life between when we dated and now. Wayne used to come to the Hat every weekend and we used to go out for supper and do this and that – now we never leave home. We always are busy doing something, and then we're tired. When you're tired, you're not interested in the prospect of packing yourself up and going seventeen miles to town.

With ranch decisions, pretty much, Wayne decides. He does ask, but in the real world I don't actually know very much, so unless it seems like a really stupid idea, I go along with whatever he says. In the day-to-day decisions, he decides. He knows what he's doing, and what he wants. I would be a glorified hired man. I just don't get paid. Longer hours, I would say, but the same, you know – he tells me what to do and I go and do it. I am involved in the goals and decision-making. He will always ask my opinion and a lot of the time, I don't really know, though there have been things I have said no to. He would like to have purebreds, and I say, "No, we're not doing that!" We sell replacement heifers as calves and he would kind of like to keep them longer, and I say, "No, this is working, we're not changing."

Being pregnant was an easy thing for me. I was twenty-six when I had Leigh. And the day before I had him, I was outside. He was born in December, and I was checking in the calf pen, doing chores. I had my coveralls on and they didn't zip up, so I had a feather coat over them. I breastfed my babies. I just thought that was the right thing to do. It's the only thing that made sense to me. We did make compromises in our day-to-day work after we had children. When the kids were little, I wasn't out very much, but as soon as they went off to school, I was out all the time. When I was outside they stayed in the house, or they came with us. We used to drag them to the barn in calving time. There's places in the barn where they could sit and watch and play. Sometimes, when they were at that unmanageable stage, they went over to stay with Grandpa and Grandma, just a mile away.

The children are involved in ranching. My kids seem very old – they know a lot and have a lot to do compared to some kids you see in town. Leigh probably drives the tractor better than I do. Actually, if I drove the tractor as much as he does, I would be as good, but I don't. If he wants to take the tractor, then go ahead, I can get the gate. Leigh is outside all the time – he likes being outside, he doesn't like to sit – whereas my son Jordan likes to sit, but he has to go outside. They're both in 4-H – they have 4-H calves and they have to feed them themselves, and they have to bed them and do all the work with them. But now they would do everything I do. Actually on weekends, when they get home, I get to stay inside and hopefully get housework done, catch up, after that whole week of it running away on me. But I don't always get to stay inside – there's often weekends where no one's in the house.

The kids are involved in some decision-making on the ranch. They have purebred cows – this is the way Wayne wangled these purebred cows in here. They've got purebreds for 4-H calves, heifers. So I guess they're decisional in that. As a general rule, they're about the same as I am – they get asked.

I'm a 4-H Junior Beef leader and I've worked on the District 10 Board in Maple Creek. Sometimes I don't do as much as I would like to do with my 4-H kids. We have record books that we go over, and we've taken tours and trips and that kind of thing. I do showmanship and grooming classes with them and show them how to get it done. Not that I would know, but I sure try hard!

I like being outside. This is a fulltime commitment no matter what the weather. I used to say that I didn't do as much in the summer 'cause I tried not to have tractor skills. I've lost that as well. So winters used to be harder because I was out more. Fencing was probably my least favourite thing, but I had to fix the fence the other day by myself and I got it up there fine, so you know things are always easier when you know how to do them. I try to avoid things I'm not very good at. I'm pretty slow with the tractor and unloader, but that's just because I choose not to do that. I'm the "gate girl." I like calving, and working with the cattle is good. Wayne hates farming, so I have learned to hate that as well. I used to like to be at that blissful place where you just said, "I don't know how to do that." But now I haven't any excuses anymore.

There was an interesting incident when we were calving here last week and I got hit in the head. The cow kicked a panel, and I was not watching and it hit me in the head and sent me across the room. Wayne thought it was tragic. It was actually bleeding quite badly and he phoned his mom and she ran over, but it's fine. I fell over and got my arm stepped on once, by a cow, but I've never been badly hurt.

I'm actually pretty good at driving. I never have found it all that difficult backing up with the truck and trailer. People still look when they see a woman backing something up and wonder, "Are you sure you can back that up?" Yes, I can.

I can see a woman going off the ranch to work if she doesn't work outside. If I didn't work outside I'd have to find something else to do. Most of the women in my age group either work in town or on the ranch. I think in order to make enough money to keep going, you have to either be the hired man or, if you're not big enough for that, have the job in town to make ends meet. If I wasn't outside, Wayne would have to have a hired man. I don't see that getting any better. I think women are either going to have to be in one place or the other. I don't think agriculture in the near future is going to be that one man working on a place is going to keep you going. And to be able to maintain ranching I think you've got to have two jobs of some kind. My husband prefers to have me at home because he can't cook and clean. Does he help with the indoor work? No! I'm always telling my children, "You are not going to grow up like that – we're going to fix that in you!"

I assume I'm not a farm person, so I don't know anything about that, but farming seems to be a very bad occupation at this point. I think we might get into trouble with farm people getting into cattle and not knowing enough about cattle and having a big discrepancy between the people who know, who have good quality end, and those who won't be profitable because they don't know enough. And you could get a lot of cattle that aren't supposed to be bred, that should be feedlot cattle. We have sold some replacement heifers to farmers who are trying to start cow herds – we work with them to get them on some kind of profitable program. You have to look at this as a business, it's the only way you can make it go. They could be good farmers, but you know, being good in one doesn't necessarily make you good in the other just because you live out here.

We don't do a lot of leisure activities. It would depend more on the year. Last year we only went out with our friends three times to golf, but the year before, we went out probably fifteen or twenty times. It depends on the difference in the year and what's happening around home, whether or not haying is going well. Social activities come second to everything else. We have some friends in town that have "Happy Hour" – we never get to it, but we would like to. Periodically we'll go to Medicine

Hat with a couple of friends who have kids the same age as ours and stay over and do the waterslide thing, have a family weekend. We didn't do that last year, but we try to get away once or twice a year with friends.

We organize a local replacement heifer sale every spring. We start planning in November and phoning people, and this goes on all the way to March. We try to find people for the heifer sale and organize how many are coming, and we have extra money for advertising and we have to decide where that goes. We make an effort to phone everybody who was at the sale last year and everybody who bought something, and find out if they're happy. Wayne spends months phoning people, whereas I organize all the background. There is a lot of public relations activity and getting sponsors as well – it's a big job.

In the summer, we seem to be busy with 4-H. The show is in the summer, and Wayne is on the Regional Fair Board in Maple Creek and that takes quite a bit of work to get everything organized in town. The kids seem to get too many animals. You know, we could cut back any time soon. But they seem to be quite happy with that whole thing. We take their animals to Piapot and Maple Creek. Maybe this year, there's a big Angus show in Swift Current in July, and Leigh has a heifer he might take down. He would like to do that, but then again we get back to this – do we really have time? One of us would have to go with him. We will see what the summer brings. Leigh likes to go and help people with cattle, he really likes cattle and he likes showing. He's got friends that are Angus people that show at Agribition and Medicine Hat Bull Show, and he goes and helps them for a couple of days. We try to get up to that show, at least.

This ranch is where Wayne says he's retiring, but I do not believe that. He says he's going to let the kids take over, but he's such a decision-making kind of guy that I don't think anyone would ever do it right. Not that that's a bad thing, because he's good at what he does, but it's an awful hard thing to follow. He says he's turning the reins over to the kids, but I don't know.

I certainly do like it out here. It's a busier life than I would have thought. You never see people working when you're driving down the Number One highway, so you don't think of it as a high-intensive working-all-the-time kind of job. It certainly is that, but I like it – the work's good for me.

DENA WEISS

*As to the viability of ranching in the future,
I sure hope so.*

Dena Weiss (Doris Bircham's daughter), born in 1964, is an accountant and manager of the Meyers Norris Penny office in Maple Creek, Saskatchewan. She ranches with her husband and family eighteen miles southwest of town and travels the Fort Walsh Road to work each day. This interview took place 33 days after the announcement that one Alberta cow was diagnosed as having BSE. Immediately, numerous countries, including the U.S.A., closed their borders to Canadian beef.

I GREW UP ON A RANCH and went to school in Piapot for the first eight years then took grades nine to twelve in Maple Creek. From there I went to Saskatoon for four years to the University of Saskatchewan, got my Bachelor of Commerce, and then moved back to Maple Creek.

After two years of university I got married, then completed my last two years. Actually my husband, Bryce, had a short stint at university taking voc ag. He enjoyed it. It was really

DENA WEISS, ACCOUNTANT, MAPLE CREEK, 2003

good for him but he couldn't wait to get back home to the ranch. So there was only one year when I was actually alone up there. And I lived with my sister-in-law so it was fine.

After I received my Bachelor of Commerce, I needed two years of work experience, so I worked and then wrote three comprehensive finals to get my certification as a Certified Management Accountant.

My husband ranches with his brother and they also have their retired parents living on the ranch. We have two children, a boy that is almost fifteen and a girl that just turned twelve. Our son, Clark, doesn't have any off-farm work so in the summer, evenings, and weekends he works on the ranch. And Shelby doesn't do a lot of work yet, but she helps whenever possible.

I stayed home with them for about four to six months and then went back to work. I was lucky enough to find some really good babysitters and I had about three main ones over the course of six to eight years. Generally they babysat a smaller number of children and that worked very well – they were all in Maple Creek. Once the kids got old enough they stayed home with their Dad when they could.

I think a ranch is a really good place to raise a family. It's like any family business – when your kids work in a family business, I think they learn economics at an earlier age. They understand where the money comes from, how you make the money. You know, you have to sell the cattle to get the money, and you have to buy inputs to go into the production of everything, and I think when kids are involved in a business such as that, whether it be farming or the local restaurant, then they have a greater understanding of economics because I think kids nowadays have a hard time figuring out where money comes from.

In the winter we spend a lot of time at either the curling rink or the skating rink. Both kids play hockey so we spend a lot of our leisure time with them, and also my daughter and I curl. In the summer we like to golf, so we fit in as much recreation as possible. I am currently treasurer of the Maple Creek United Church and the Maple Creek Arena Board, and a co-coordinator for the Maple Creek Cattlemen's Golf Tournament.

I spend my holidays working on the ranch mostly. When there's something that needs to be done with the cattle, I help with that – riding and sorting and preg checking and all those sorts of jobs that need extra help. I enjoy riding and helping. My husband's brother doesn't ride much, as his expertise lies more with the farming, haying end of things.

I have worked as an accountant for seventeen years. I would say seventy-five percent of my clients are ranch people and the other twenty-five are town or business people. Every situation is different. I have situations where I've never met the husband, where the wife does everything. But I've also had situations where the husband does everything. It's actually almost rare that I see both at the same time. There's usually one of them that takes the initiative and is the bookkeeper.

As to the viability of ranching in the future, I sure hope so. I guess the trend that everybody sees, not just me, is the trend to larger operations, declining margins, necessitating an increased size of operation to get the same return, so that's sort of discouraging but it is a trend. The margin of net income tends to go in cycles, too, just as we have cycles of good years and then we have bad years.

I would say financially, the issue of BSE hasn't shown up in my business yet, but I certainly get a lot of phone calls and when the NISA [Net Income Stabilization Account] options notice comes around and there's an option to get that money out, people aren't hesitating to get their hands on whatever money they can. They're bringing in that new NISA program so NISA accounts are going to be collapsed, but I would say they are probably going to be collapsed sooner rather than later because the government will give you five years to do it and I don't think people are too worried about waiting five years to collapse them to spread out the tax. They are worried about getting their hands on the money now because they need it to operate and they're scared.

I would say it's impossible to predict right now how much impact BSE will have. It could be very wide-reaching, depending on how quickly the border opens up and how quickly the price stabilizes. Right now we're dealing with people wanting to market dry cows and cull cows, and of course, there are people with fat cattle. I would say that come fall when everybody is marketing their calves, that's when we'll know – if the price is half of what it was last fall, then everybody's in big trouble.

I guess I feel I do have a little bit of farm/ranch knowledge heading into this job, and I think that tends to help me understand the problems that farmers and ranchers are facing because, you know, they're very real and they're perhaps easier for me to understand but that's not to say somebody not from a farm/ranch background can't understand these. It's maybe just a little more difficult – even the terminology is difficult. I work with somebody who is from the town who's never had anything to do with farming and ranching, and she does very well because she asks a million questions and she can learn the terminology. She picks things up, but it's certainly more difficult for her. When it comes to cattle, for example – it's interesting when terms come up that can be taken completely out of context, and we all know what they are but an outsider has no idea.

Women of my generation are certainly more visible in the workforce off the farm or ranch than they were forty years ago. Choice or necessity? I would say that it's a little bit of both. You know, sometimes choice becomes necessity because you get accustomed to the extra income or it just becomes routine – it's sometimes hard to say which came first.

The initial shock of finding a cow infected with bovine spongiform encephalopathy in Canada has now worn off. Ranchers on the whole are discouraged. I see some extreme pessimism out there, but they are trying to hang on. Many cattle producers are keeping their cull cows and continuing to breed them, hoping for the United States border to reopen. Even if that happened immediately, it probably wouldn't help our cow markets until far into the future. At this point we can't ship any live animals across the border no matter what age. Some ranchers are unwilling to sell into a depressed market. Others are taking their losses and marketing their cull cows for anywhere from three cents to twenty cents a pound. Before BSE these animals were bringing in the neighbourhood of forty to sixty cents per pound. Following the BSE announcement on May 20th, 2003, cattle prices were literally cut in half. Considering all groups of cattle under thirty months of age, we have now levelled out to about a twenty percent deficit.

I have reports of people shooting their cull cows and cull bulls. These animals still qualify for BSE testing. Canadians are now attempting to rebuild a solid infrastructure so we can handle mature animals, but this takes time and lots of capital. There is no doubt that our cattle inventories are increasing, especially in regard to mature animals.

I haven't seen a lot of numbers for 2004, but calf prices in the fall of 2003 were relatively good. This fell apart when a second case of BSE was found in Washington on December 23, 2003, and the infected animal was later connected to an Alberta dairy farm. Returns for 2004 will most likely be down. The cow-calf producers who sell calves right off the cows have probably been affected the least, while those ranchers who retain ownership until their cattle are finished are taking one of the hardest hits.

Politics and science are miles apart. I see a mistrust of government. Many of the ad hoc programs to address the BSE problem have no rhyme nor reason and at best are only band-aid solutions. Banks are naturally becoming concerned. Operating loans are growing. Many in the industry are working tirelessly to make us less dependent upon the United States, to establish more markets, and to strengthen our industry internally.

LOU FORSAITH HALTER BREAKING THE BABES, 1930s

LOU FORSAITH

The only thing I can tell you is enjoy your life on the ranch while you're there, just enjoy.

Lou (Wetherelt) Forsaith was raised with her five sisters and brother on a ranch in the southeast corner of Alberta at Onefour. She started school there in 1920, before her fifth birthday. In 1940 she married Stuart Forsaith and in 1949 she started ranching with Stuart and their daughter Robin (Wolfater) in the East Block of the Cypress Hills, south of Carmichael, Saskatchewan. Her daughter Lyn (Sauder) continues to ranch there.

MY FATHER MADE MONEY by freighting for the ranchers when they first went to southern Alberta. So that left mother alone, at first in a tent, with the three girls. You know, they had to break up so much land to prove the homestead, so she did the breaking, and with these three children – the milk cow babysat them. The kids followed until they were tired. They'd lie down and the old milk cow would lie down with them, and she'd stay with them and mother would go on working right in the same field. I suppose the youngest would be two or three and the older one would be, I'll say, five, six, probably.

And Dad would be away because they were hauling supplies from the States, from Guilford, Montana, and from ... I think Foremost finally came into the picture, but originally it was Seven Persons, and they hauled supplies from there. Mother had the kids and four horses and the milk cow, and was breaking up the land. So I suppose any decisions made were probably mutual – they'd have to be, wouldn't they?

Later, the older girls ... of course, times were tough then, and they worked whenever they could, wherever they could, like cooking for haying crews and that sort of thing. And I took the place of the boy they didn't have, so I had to get out and scratch. I had to learn to drive horses – two, four and six. And I had to look after them – well, I had to do what a boy would have done. There are three older sisters, then there's a seven-year gap and I'm next, the oldest one of the younger four. Now, you didn't tell any stories in our family and if you said you were going to do something, you did it. Just like they sealed a deal with a handshake, if you gave your word you kept it.

Before I even started school they had a big Christmas concert and I can remember bits of that, and there must have been a basket social, you know what that is – every lady took a basket with a lunch in it and then the men bid on it. The money that was taken in from selling the baskets was put in a fund. Anything we needed, that supplied. Now this goes back ... see, some of those kids didn't have money enough. We got our pencils and our scribblers, I think maybe we got our own slates, but everything else was bought for us until, oh, I must have been in grade eight and then that fund ran out.

Schooling was kind of a here-and-there situation. We didn't have a proper schoolhouse. We went to school in whichever vacant house was central. The first school I went to was seven miles from home and it was a vacant house. Then the people that were the farthest to the north of us moved away. Well, then we went to school down closer to home. But we at that time only had school during the summer, because in the wintertime, you know you couldn't heat those old places. And it was cold getting to school, too, and those kids were all ages. Then one year my younger sister and I went to school in Montana. I was in grade eight and before I could write my grade eight exams down there, I had to write grade seven exams, so I wrote my grade seven and then I wrote my grade eight, to get high school entrance. Then when I came back to Alberta they didn't recognize U.S. grade eight, so I had to take grade eight again.

There were very few children at school. My mother looked after a couple of my cousins and they went to school with us sometimes. And I don't know what you'd say the others did. I think about it now, and I think they must have nearly starved to death because I think the fathers were kind of useless. I do, I think they must have lived on welfare. There was nothing wrong with the children, mind you, but poor little beggars, you know, it was kind of rough on them.

Now, I was kind of a boob, I think. When my older sisters would be going to school with this Miss Smith – she came from Nova Scotia – apparently I'd howl and bawl I wanted to go to school. So the teacher said to my mother, "Send her to school for a week and then you'll hear no more of this." Well, at the end of the week the teacher said to mother, "Just leave her alone, she's doing fine." So I wasn't even five when I started.

There were so few of us, there were probably only one or two boys in the school. We did whatever – if we were playing Ante-I-Over we all played, we all played duck on the rock, it didn't make much difference. There wasn't too much baseball in the earlier grades because I don't think they had money enough to buy a softball. But when the school moved down a little closer to home, things must have been a little better because we got a softball and a bat. And those kids were taught to behave at home and I don't think the teacher had trouble with them. When you were old enough to understand, you knew to behave.

I finally went to business college. If I had been a boy, I would have gone on to do something, probably working for somebody else with cows. When Stu was overseas, one of the dealers came to me and wanted me to take over the management of this ranch in the Cypress Hills. Well, Stu was coming home, I wasn't going to be working, you know, I couldn't. That would have been the only time when it was really worth money. I didn't do it, but I had the opportunity. Before the war, the day you got married you finished work. Married women did not work. Then somehow or other we got tangled up in that war and then you could work, and so, when Stu was gone, when I went back to Swift Current, I worked there. Then an opening came at the experimental station at Manyberries, so I came back there, and I worked there until he came home from overseas, which was about three years.

Roberta was born in Minneapolis-St. Paul, where Stu was finishing his master's degree. He was an agricultural engineer. He had started his master's degree and then the war came, and of course, he wanted to go. So then when he came back, he finished. Well, of course, you always finish something that you start, so we went down there. And Lyn was born in Gull Lake, after we started ranching. Part of our reasoning to go to a ranch, or to get away from the city, was we wanted children, but we wanted them to grow up with a sense of responsibility. We wanted them to have a better viewpoint than we were seeing in children raised in cities and small towns. And I think it worked. We wanted two children, but one of them was supposed to be a boy! But it didn't matter. You know, you wanted a boy to turn the ranch over to when you were finished with it. Well, we got two girls and they both wanted to ranch! They were given an opportunity to choose what they wanted to do and one is a dietician and the other is an occupational therapist, both loving the ranch. Their future will be, half the ranch goes one way and half the ranch goes the other way, because they love it.

I worked at whatever needed to be done. Well, of course, that was horsepower – I worked with horses, broke a few, enjoyed it. In the spring, when the weather was good, I enjoyed being outdoors. But I liked the riding and the cattle better than the fieldwork. I wasn't too enthusiastic about that.

I usually had help in the summer during haying time because then we were working up on the bench. Although with Robin, I often took her up there and she would sleep in the shade of the stack. I used to worry about this, I thought, "What am I doing to this little girl?" And when she got old enough, she told me that that was a very happy time for her.

Lyn's an August baby. It was getting on to winter when Stu needed somebody to drive the team while he was doing something with the cows. So she'd probably be five or six months old. We'd roll her in a quilt and put her between two bales on the hayrack, so she couldn't fall off, and you know, she never howled – she liked that. Someone had to take the little ones out – if there was a storm and the children were small, you didn't leave them alone in case of fire. The little ones would look at the rabbits and chickens and things.

You know, when you live on a ranch and you're isolated in the winter, you get cabin fever a little if you don't move around. So sometimes Stu would put the supper on and watch the kids and I'd go and do chores or something, I'd get away from the house for a little while. He was quite willing to do that.

Delores Noreen and I had a common interest in horses. Oh, this was a bad winter and snow like you wouldn't believe. So I went over to Delores and Duff's and we rode to Tompkins, fifteen miles, to get some cigarettes, I think. Oh, that really wasn't … I mean, there was more to it than that. I visited my sister who lived in there. Delores had rolled some cigarettes and put them in her pack. So we stopped on the way in to have a cigarette, but when she opened the pack, these were homemade cigarettes, and the tobacco had all shaken out, so we didn't have a smoke. Then when we came back, Duff didn't want me to go home alone because there was so much snow. Well, we got into snow that the horses couldn't manage, you know. It was right up to their bellies. So, I got off my horse and the snowbanks would carry me, so when the horse didn't have me to carry she could plunge through. But oh, we had more laughs about that because we were stuck!

When Stu's health failed, life on the ranch changed because I couldn't leave him alone. I would say, "Don't try to go downstairs, now, I'm going to run to the garden." And I would go to the garden, but every fifteen minutes I would come back to check that he was all right, and nine times out of ten, he would be trying something. And you couldn't scold him – I mean, he didn't care much for being tied to a wheelchair. I tried to be patient, and I think I was quite patient with him. I would take him out to his shop, but he couldn't do anything much. Then I would try to do what he'd tell me to do, and my fingers were all thumbs when it came to woodwork, so yes, it did change. And if I wanted to spend a day riding, someone had to be with him. I couldn't leave him. As soon as his health began to fail, there were other changes. Then I was the one that was making the decisions, financial decisions and the whole bit. Eventually we moved into town.

I've watched my grandchildren grow. I'm very proud of Robin, you know, she has brought those children up to accept responsibility. If there's a job to be done, they do it without being told. They can see for themselves.

Ranching used to be a way of life, and now it's become a business. And it takes a lot of money to keep it going, it takes some very careful management, but money is the main thing

– if you've got enough money you can get away with fairly poor management. I'm not too sure that women today take an active part in the ranch itself, because they are away working, trying to make a dollar to hold the ranch together. I don't think we'll ever get back to the early way of life, I think it will always be as it is now, a business. Not too much joy in it, it'll just be a money-making scheme. It'll be like farming. You know, farming is not what it used to be – now it's get out and go and make the money. And I'm afraid ranching is going that way, too. It's a pity because, you know, it was a very pleasant way of living and I don't know if the children will ever have an opportunity to see that.

Lou Forsaith passed away in the fall of 2004.

ROBIN WOLFATER WITH MURPHY, 2003

ROBIN WOLFATER

Well, I guess we kind of enjoy working together – he needed my input, he was a farmer and I was more the rancher type when we bought this place. So that worked out.

Robin Wolfater, born in 1947, is the daughter of Lou Forsaith and the sister of Lyn Sauder. Robin married Ronnie Wolfater in 1970. They have three children: Rajanne (Wills), Renee, and Bradley. The Wolfater ranch is located on the north slope of the Cypress Hills, along Skull Creek, in southwest Saskatchewan. Robin completed her degree in home economics at the University of Saskatchewan in 1969.

WE FARM AS WELL AS RANCH – you kind of have to do both things to make a living. I like the cattle, that's my thing. I like helping when they're processing them and checking to see if they're sick and helping at calving time, riding, counting, making sure they're all accounted for. I'll drive the tractor to square bale, but other than that I'm not into machinery, although moving machinery isn't bad. Ronnie and I do books together. I get a little pressure that maybe I should do them myself, but I think that if it's a bill, I want to know what it is, and where it should go, so we just have always done our books together.

Our son, Bradley, works with us. He bought a quarter of land last summer and you know, we need the help – we have to do more and more to make a living it seems, so we're kind of hoping to work into it. I don't know how that's going to work when there's income for one family and work for four or five. Physically, I'm not as strong as I used to be, so I'm doing less whether I like it or not. As we approach retirement, I guess I'll still be involved in management. I still check cattle, but I maybe can't pull a cold calf out of the creek or something, like I used to be able to do. Bradley is doing more of the rough stuff. I don't take chances on my horse like I used to – I like to do the slow pace and not end up bouncing.

We discuss and make decisions together. I'm sort of trying to back off, especially with machinery things or day-to-day things. I think Ronnie and Bradley should talk about those things. I think if that's the way things are going, I'd better be pulling out of it. It was kind of hard the first year or two. We talk about our goals from time to time. We really don't have anything written down from year to year, but we talk about where we'd like to get to. Bradley has some cows here, and we're trying to get him thinking about what the costs are, and Ronnie likes to sit down and do a break even if he's going to buy some calves. So he can do that.

I think if we continued to live here, on the ranch, we wouldn't really be retired because you'd still be stressing over all the things that are going on. I suppose for peace of mind we should probably move away, but I don't know, I can't see it in the near future, for financial reasons. Because Bradley, you can't expect him to pay the going price for land, he can't, and I don't believe in giving him the land. It's something we've certainly got to get planning so everybody knows where they're at. We have two daughters. Rajanne and her husband are about fifteen miles from here – they bought a small place. So we help each other quite a bit. Renee is a nurse.

We wanted our children to go away and get some kind of education. You know, get away from home for a bit. Now, I've told Bradley different times, "Don't get tied to the ranch." Because he might marry someone who wants to live in Timbuktu, and he should try to keep his options open. I don't know, I think he's got his heart set on it, but there's life besides this, and it may be easier. It's probably easier to say these things when this isn't the Wolfater or my home place. If we were to sell out tomorrow, there wouldn't be anybody saying, "Oh, you sold my home!"

Mom and Dad were always quite adamant that my sister and I get some education, so if we were ever left alone with kids, we could look after ourselves. I took my degree in home economics and my dietetic internship. I suppose some of that was from thinking that someday I would make a good mother, or make a home. Certainly there weren't jobs for dieticians in this area, but it was something that I was interested in. If I was to do it again, I think I would like to do environmental studies and biology. I really like biology. But that's in hindsight. I didn't actually think I would work in this profession because Ronnie was in the picture then and there weren't any jobs right here. When my friend in Swift Current was on maternity leave, I worked part time for three months. Rajanne was just two. But driving down there wasn't much fun. I also took a position for four or five years in Gull Lake Hospital as a casual cook. Money was short, and so it wasn't a lot of hours a week, but they were twelve-hour days, so when you got there the day's pay was worth going. It was a seventy-mile round trip.

In the future, I think most ranch women will be equal working, business, decision-making types – it'll be just a necessary thing. Most of the time when women work off the ranch it's a financial need, and I think it's unfortunate that these places can't make us a decent living. The children, I think they need a mom, everybody's happier when mom's home. I'm sure they get satisfaction from working away, but then they're tired when they get home. If that's what they want to do, great. It's just too bad that things weren't like thirty, forty years ago when there was enough income that you didn't have to think about that.

We women have to get outside and work on the ranch because we can't afford hired help. When my mom was my age, that was just part of it, you had a hired man. And even when she grew up, they had a hired man around to help with some of these things. And at that time, too, men would stay the winter, just for room and board, and now, I mean, certainly their wage should compare with what other occupations are getting but we just can't afford it, so we women do have to get out and help. Some of the changes are for the better – I think it makes the family work together more, but when you have to do more and more to try to make a living it gets pretty stressful. I guess we have to learn how to deal with the stress. I think I had an idea that maybe by the time we were fifty that we'd have reached an easy time of our life, but really it hasn't happened.

Certainly I've had to cope with some health-related problems, and they can affect your ability to ranch. I tend to keep things inside, and it brews and doesn't do you a lot of good. I suppose I've had times when I've been depressed, but I'm sure not going to go out and tell anybody. I found it pretty hard when Dad moved into the nursing home. I suppose that was kind of a start of the grieving process. You'd come home from visiting there and you'd be really down for three days. That took quite a while to get used to, and I'd feel guilty about not spending more time with Mom.

We live in the Skull Creek area – there are several families around here and we have a little community centre. It was

Mannville School. It was moved from about four miles east. There was a bit of controversy over that, but it was felt it would be used more if it was beside the rink. And it is used for 4-H meetings and speeches, and by the light horse group. There used to be card games there, too, but some of those people have moved to town so that doesn't happen so much. They have a social there after the bonspiel. Especially in the wintertime, it was a great place to get together. In the sixties there were nineteen rinks in a square draw. People would come maybe ten, fifteen miles to curl, but then, there were that many people in the area. We still have the bonspiel about the second week in January, but everyone's so busy, and with the kids grown up, they really don't even get little square draws going anymore. Over Christmas, this year, it was so warm it was a panic to get the ice in before the bonspiel, but it happened. And kids have skated there, on the outdoor rink, this year. Well, maybe they only got a couple of weeks of skating, but they made the ice anyway.

Having babies complicated things for us a little bit because when we moved up here, Rajanne was two. She could travel around in the truck with Ronnie, and I could still ride, but the poor kid had to be outside quite a bit. I think that's why she likes being outside now. When Renee was born, she was old enough in the spring, she'd ride around in the truck and be quite happy, but when Bradley was a baby, all he'd do was bawl in the truck, so I couldn't get out that spring and do my riding. That was hard. Then we hired Beth Hobbs in the spring, when she was finished school in Red Deer. She came for a good six weeks so that I could help fence and work with the cattle, and it was much easier.

I have done some volunteer work. I enjoyed my stint on the Home Care Board in Maple Creek. I've helped with 4-H in Tompkins and a little bit down here, too. I helped with the Skull Creek Curling Club Ladies – they don't do that so much anymore, but if they were putting on a supper, I mean, you did your part. But it isn't something that I do three or four times a week.

I've always been interested in art, and I've taken the odd workshop or class. I like sketching and I've been trying watercolour, but you just have to take some time to do it. I find that kind of hard, especially in the house, because people are coming and going, the phone's ringing, and you just can't get into it. But I've been a member of the Art Club in Maple Creek for a couple of years – maybe you just go and visit with the women, but it's a reason to get out of here and we're doing a few things. We're going to do a calendar for Maple Creek's hundredth anniversary next summer. So it's nice to have a group to do things with and I enjoy it.

Sometimes, when it's the high workload period, there isn't any leisure time. I don't do a lot of curling, but just this last week I was curling with the other women. Ronnie and I tend to like nature things, I suppose because we're exposed to that so much. We're trying to do a little golfing but don't get a lot of time to practice. It's about thirty miles to the Eastend course. Some of the women have asked me to go in and golf, but I just don't feel I can pack up and leave when everyone else is busy. We're members of the Saskatchewan Stock Growers Association. I think that these cattle-related organizations are recognizing women's contributions more. The president of the SSGA is a woman now, Marilyn Jahnke. I still think they tend to be a man's organization, but there are good strong women involved – they know what they're doing.

We've become involved with the Native Plant Society of Saskatchewan, NPSS, partly because we're noticing all the native seed on the grasses in good years and there's a sale for it, for reclamation uses. I'm interested in the flower seeds, things that we have here. There's a market for some of that and I'm actually a director of that society now, but I don't feel

I've been giving them a lot of help. I don't know if I'll continue on, but I really enjoy those people and going to their annual meetings and things.

We felt honoured to receive the TESA [The Environmental Stewardship Award] in 1994. The Saskatchewan Stock Growers wanted to try to make urban people aware that we really do care for and look after our land, like "we" as in ranchers. And this was one way to get a little publicity. I don't think we did anything any better than anyone else, but I guess it was the first award given out for environmental stewardship. People came and we showed them around and they asked us questions about what we did and it was kind of interesting.

I don't know if it would be my philosophy, but I think you develop a love for the great outdoors and nature, so hopefully we get respected for that – and try to preserve it, look after our environment. Because I think of our future, our grandchildren, or the ones coming after us. Southern Saskatchewan is one of the few places in the world where there's some good fresh water and air. It kind of makes me shudder to think of what may be coming. Let's look after what we have and enjoy it.

RAJANNE WILLS ON OJ, AND LYN SAUDER ON JAZZ

LYN SAUDER

We're not raising cattle anymore, we're raising food.

Lyn Sauder, Lou Forsaith's daughter, was born in 1952. She ranches on the Forsaith home place with her husband, Glen. Their ranch is located on Bone Creek in a valley banked by hills that rise over 3,000 feet above sea level. To the south is the bench where much of their haying takes place. Lyn was raised here – she is a rancher – but she is also an occupational therapist. She travels throughout southwest Saskatchewan, four days a week, with her off-ranch employment.

AS WE WERE GROWING UP, we were always involved in everything that was going on. I remember when Dad would hire help just for haying time – I also remember when he started having hired help year-round. When the hired help was here, we probably weren't as involved with particular chores. I didn't do a lot in the hayfield compared to some girls growing up on the ranch, but we were always involved.

Brandings and roundups were scheduled for when we were off school, on weekends, and I guess I felt kind of important in that respect. As a kid growing up, before I actually got moved back to the ranch, where now I'm financially responsible, I always remember those jobs as being fun. But, of course, I didn't have any of the responsibility or any of the expense connected to it. Now we look at it in just a little different light. I can remember when Delores Noreen used to come and rope for us at brandings, and Lawrence Theil and old Alan Bowes coming down, and he'd tend the fire and you know, it was fun. Back when we were doing roping the old way, there would have been twenty-five people, I suppose. Then when we had to go to the calf table, which quite annoyed me – of course at that time I also wasn't doing the wrestling – but then our branding crew became a little smaller. The meal preparation was definitely the major thing for Mom.

I went to Alpine School for three years – it was about three miles north and we rode to school. Eventually, when our neighbour was driving school bus, it ended up in our yard, so that was quite a change. I can remember – of course, I was eight years old – not being very happy that I couldn't ride to school anymore. We were friends with pretty well everybody at school. Lots of our activities related to horseback riding. Summers we'd get together and ride. When our neighbour east of here was going to school, we would ride home together, and we would have an odd horse race and get dumped off once in a while in the spring, when the horses were feeling good. I spent most of my summers riding, checking cows, and swimming in the creek.

I went to Gull Lake for high school, twenty-five miles away. I didn't participate in any extra-curricular sports – I guess my one activity was 4-H. I've been involved in 4-H since I was old enough to walk, indirectly, and then later as a member. And that was a big part of our social life in a sense too because, I don't know if it was every week, but it was fairly frequently that we got away for something to do with 4-H. And that's something I liked and stayed with, but I kind of wish I'd gotten into band or something else, too. Now I think it was maybe a good thing not to be so involved, but at the time I remember feeling very left out. When I got into Gull Lake I got to be friends with some of the girls from town. I remember my first, I think actually my only Halloween, going trick or treating with one of my friends in town – I'd never seen so much candy in my entire life!

I remember my parents telling me almost from the time I was old enough to remember that I was going to get some additional training. I wasn't just going to get grade twelve and sit at home. And there was a point in time, in high school, that I would have happily quit. My parents said, "Well, I guess if you're going to quit, then you'll have to get a job, but it can't be here." "Well, then, I guess I'll keep going to school!" I thought it would be just kind of nice to stay home. Anyway, we were going to get an education, but the choice was ours. They would say, "If you never use it, fine, you at least have the training. You have something to fall back on if something should happen in your life that would make it necessary to work out of home."

I went to the University of Saskatchewan for my arts, and then I applied for occupational therapy and went to the University of Alberta in Edmonton. I used to really enjoy doing crafts. We'd taken leatherwork and I used to do lots of "arts" types of things. Mom and Dad – if there was anything we expressed an interest in, they would always make sure we got something associated with it so we could at least give it a try. So entering university, I really didn't know what I wanted to do, but I knew I didn't want to be a nurse or a teacher. So reading through a career booklet, they had a description of various occupations and occupational therapists were apparently people who used crafts in the treatment of patients with physical and psychiatric disorders, and I thought, "Wow!

What could be better?" And now, it's just totally different from that. They used to sit and make baskets or weave and make pottery, and so I remember doing all those things in our training, which they do not do anymore – they just learn how to analyze the activities. But I can remember sitting there at the wheel making pottery, thinking, "If I had to do this as a patient, somebody's head would get smashed! This is so frustrating!" Now, yes, they will do activities, but it's certainly more like checkerboards using Velcro and reaching up a wall using a loom to increase shoulder mobility, for example.

I'm working, not quite full-time, but I work basically four days a week. I'm not as involved as I'd like to be with the cattle, but I still assist with that on weekends. I used to spend hours riding, when we were children of course, but I don't get to do that now, which I miss. But you can only fit so much into a day. I'm still involved in the overall decisions – I mean, Glen has to make the day-to-day decisions because I'm not here, but he doesn't go off and make decisions without me at least knowing about what's happening, which is very nice. We still discuss the major decisions, you know, if he's going to buy a new piece of equipment. He makes the final decision because I really don't know anything about buying a tractor, but we still discuss it, and the financial implications. I think sometimes Glen thinks that if I didn't work off the ranch I'd be home to be helping more, but he also realizes the importance of it financially. So I'm away at education things from time to time, and he doesn't really begrudge it at all. I think he realizes that I have to keep up with what I'm doing. You know, I would say it's a positive attitude. It was certainly different for Mom and Dad – I think they discussed pretty much every morning what they were going to do for the day, kind of made decisions, then of course, made the major decisions as well.

Actually, Glen does help me with the indoor work quite a bit. In winter he does probably most of the suppers, or part of most of the suppers. And he helps, not much with cleaning the house, but he's not the kind of guy that you have to be picking up after all the time. He'll do the dishes, especially if they're sitting there two or three days. If I leave the dishes long enough, the "maid" will get to them!

I think now probably more women work off the ranch than did when I was a child, just out of need. I guess if you're career-minded, obviously there have to be adjustments when you work away from the place – someone has to take up the slack. But if you are career-oriented, maybe it is more satisfying for you, in that sense, but I think it does take away from the ranch operation when both aren't there.

Our goals evolve out of conversation. If there's something that we really want to do, we'll discuss it and we'll just kind of see where it goes. We'll see financially if it fits, and timewise if it fits. If there's something we definitely want to do, then it happens. Just having recovered, or recovering from a year of drought, our goals right now are just getting finances in order and carrying on from here.

I always assumed I would be a mother, but I started thinking more seriously in my early thirties and went through some of the testing. I don't know if there was something we could have done about it or not, but I had heard of so many couples going for in vitro fertilization and spending thousands and thousands of dollars and still not having any successes. So we did consider adopting, and we got adding our ages up and thinking, at that point in time, it was at least a ten-year wait for a healthy white baby, and possibly it's longer by now. We thought, that's not fair to a child, to adopt when we're grey and rocking in our chairs. So I had never talked to Glen about this, but I thought maybe we could adopt a disabled child – then I thought, no, I can't give the time to that child that he or she would need because financially, I do have to work. And I thought, if I go to work and I come home and I don't have patience for a child, then it's not fair to him. So that's where it got to.

Probably as we age, our roles on the ranch will change out of necessity. Although I don't know that what I do, that my role is physically strenuous, but I would see Glen's role as probably having to change. In terms of health, I was diagnosed with multiple sclerosis a few years ago. And I am on a medication that seems to be maintaining – I don't know if it will. If it doesn't, I guess, just judging from other neighbours with MS and what has happened, that might very much change my role. But so far it's been working, and I've decided I'm just going to do what I want to do until or unless this starts to affect me. I used to think about retiring off the ranch and maybe doing a little travelling, but I guess as I get a little older, I think it's just nice being safe at home. I hope I can, as long as I'm physically able, stay here.

Our leisure time is pretty much settled around cattle. I always joke with Glen, I'll have to find a cow down in Mexico if I'm ever going to get him there. Agribition, of course, is an annual event for us, and that's only because there's cows there. Glen enjoys riding, but not for enjoyment's sake as much as I do, so we haven't got into any riding clubs as such, although I wouldn't mind doing that at some point in my life.

We're members of the Canadian Charolais Association, and Glen is involved as a committee member on that and on the Stock Growers. I tend to go along when he attends some of his meetings. I was on the Crisis Services Board when we were still at Rush Lake. I've been on the board for the Saskatchewan Society of Occupational Therapists, in the first years of setting up the Council structure.

The future of ranch women – and I'm just speaking very generally and this is just an opinion – but it's looking like if ranches are going to survive, they're going to have to be bigger. The place that we're on, the size is kind of in between – it's not quite big enough that we can have enough cattle to make a viable living, I guess. So, I think either places will be bigger or smaller. I think of my niece, where they have some land and some cattle and they both work off the place, but it's still manageable. They're both happy to work off the place. Or else ranches have to be big enough that you don't have to work off. And it seems, raising cattle, there's so much involved now, a decision that the American government makes can impact what we get for our cattle, overnight. And hoof-and-mouth in England can impact what happens to our cattle, overnight. I think years ago it probably still affected us, but it took much longer – now we know about it the instant it occurs. It doesn't seem to me now, but I guess that as I was growing up, it was more relaxed. Maybe that's just because I wasn't involved, I didn't have financial responsibilities, but I think now we are more politically affected. The value of our dollar alone can affect what we get for our cattle. So the whole attitude of consumers, and I don't disagree with it, but the whole food safety issue is becoming so critical, and it's starting to impact how we run our cattle. There's none of this "give them a shot wherever you can get the needle in" attitude anymore. I mean, you want to be so careful with how you're managing your cattle from the minute they're born until they're sold. Sometimes it's economic and sometimes it's just the time factor, but definitely much more consideration has to go into all your decisions.

I think the "Quality Starts Here" program that the Canadian Cattlemen's Association is trying to get into play will become a critical thing in the future. I'm not certain about everything that is involved, but I think that's going to involve keeping records of the shots that every animal gets and where they get them, just keeping accurate records of what you're doing with all your animals. I think that people who are very much opposed to this – I mean, we're a very independent breed, like "Nobody's going to come in and tell me how to run my cows!" But if we're going to be serious about raising food – I mean, we're not raising cattle anymore, we're raising food. If we're going to be serious about that, we're going to have to take on some of this, and people who are seriously opposing it, saying they're not going to do it, will eventually end up not being involved in raising food any longer, no matter how many years that's going to take to filter out. So every job means more paperwork. Sometimes it gets pretty frustrating, but when I bite into a steak, I'd like to know that it's safe.

My philosophy of life – it's part of my upbringing and part of getting the education so I had the tools to look after myself. I guess I would encourage any ranch woman, or any woman actually, to look after herself, but not to the point of neglecting husband or children. I mean, we have to be inter-dependent in the family, but don't let anybody take advantage or abuse, just the same as you should not in return. There has to be a respectful relationship in families. Just watching different families around where parents are respecting their kids, there's quite a different feeling than where they're not respected. Otherwise my philosophy in life is a very strong work ethic, but I also believe that you have to give back to community, which I'm not at this point in time but I will someday, I hope. But you have to give back to that which is making your living in some way. It's more than just staying on the ranch, feeding your cows in the winter. I mean, you have to be contributing in other ways as well. And I guess that part of looking after yourself is taking time to do things for fun. Whether it's just to sit with your neighbours and have a good laugh, or if you're into ski holidays or something like that – I mean, we're not that extravagant or don't feel we have that kind of time, but just take some time for some fun.

RAJANNE WILLS

The last two years have been very busy,
but this is where we want to be.

Rajanne (Wolfater) Wills was born in 1972. She and her husband, Grant, recently purchased a ranch on Jones Creek, twenty-five miles northeast of Eastend, Saskatchewan. They ranch together, however both Rajanne and Grant have full-time off-ranch jobs as well. Rajanne took agricultural business at Olds College. She is the daughter of Robin Wolfater and the granddaughter of Lou Forsaith.

WHEN I WAS LITTLE, I went everywhere with Mom and Dad when they were doing chores. If they had to go riding, I went riding, too. I rode behind, or Mom would ride and I'd be in the truck with Dad. Then when Renee and Bradley came along and they were quite small, Mom couldn't go out as much. When they got a little bit older, they would ride in the truck with Dad and I was old enough to ride, so Mom and I would do our own thing. When it was really busy with fencing and farming, Mom would get a babysitter to come in for the summer months, for a few years, to look after us, so she could go and help Dad more. I had regular chores. My main chore

RAJANNE WILLS AT THE CORRAL, 2003

in the house was to do dishes every night after supper, which I did not like! Or cooking, helping with cooking, which was fine, and outside, helping Dad drive truck, opening gates, chasing cows, riding, and such chores.

Skull Creek has a lot of neighbours close by, very supportive neighbours. In the winter we would have curling down at the curling rink, and there's a skating rink. That's where I learned to curl and to ice skate – I was quite involved there. There would be Christmas parties that we'd all go to down there, and sometimes barbecues in the summer. Everybody would get together – it was an active community.

I didn't really have any playmates except if friends came to visit. I was young compared to everyone else in the area, and there was no one younger than me. So my playmates were my dogs, my cats, and my horse. I think that's why I like animals so much – it seems to me that those were my friends. I spent a lot of time with them and just being by myself. I don't mind it as much – I've spent a lot of time by myself. Some of my friends weren't from a ranch, they were from grain farms, but they weren't right involved with the cattle or horses. They thought it was quite neat that I got to do all that. Where I lived was really nice, and since I had my own horse and dog and cat, they liked to come out and see them.

I went to school in Eastend. It would be, in and out of everyone's places, probably a bus ride of between thirty-five and forty miles. We got on the bus at twenty to eight, and some nights we weren't home until five o'clock, depending on who was dropped off first. I was in 4-H for quite a few years, through high school, and I was president of our high school in grade twelve.

My post-secondary education was influenced by the fact that I lived on a ranch. I went to Olds College and took agriculture business. I wanted to take something in the agriculture area, something to do with animals, and with that course I got a little bit of everything. I ended up with an agriculture/

business finance major diploma, which was geared more to working in the banking, lending area. But we also had courses in beef production, soil and plant classes, land resource classes, and accounting and marketing. Now I have two part-time jobs. I work for the Town of Eastend/RM of White Valley and for Investor's Group – I'm the assistant administrator for both.

After Grant and I finished college, we worked for six years – actually, we lived in Alberta, and we always wanted to come back to Saskatchewan to ranch, but we needed good jobs. So Grant ended up getting a job with an oil company, Renaissance Energy, and we moved back home right after we were married and started looking for a ranch in the Cypress Hills. We kept looking and we were starting to give up. Then finally, two years ago, the opportunity came up to buy the Sinclair Ranch. We were both working and we knew we were going to have to continue to do that to have our ranch. So we both still work off the ranch, and we will until it's paid for, I guess. But it takes away from the time at home, at the ranch. We're busy with work and then when we come home, we're busy fixing and cleaning and getting things organized here at the new place.

Grant gets five days off in a row, and during calving, I'll take some days off just so one of us is home. If necessary, we'll ask one of the neighbours to come over and check if something is calving during the day. Everyone's been helpful with that. We haven't lived here for very long, but the neighbours are very supportive and helpful. They were glad that a young couple moved into the community, so that's great.

Grant encourages me to come out and work, and he really appreciates it. I will help him with the mechanics but I won't go and do it myself. Everything else that we do, I do. I help feed cows, ride horses, chase cows, check cows, help with the branding, the cleaning of the barn, all the animal care. Indoors, Grant helps with the dishes and will vacuum and do laundry.

Really, only the last year, I began thinking about motherhood. I think I want to have children, but we're still discussing it. If we weren't living on a ranch and were in town, I wouldn't consider it. This lifestyle is great to grow up in.

In my leisure time I go horseback riding, often. I do that by myself, and I'm outside gardening and working in the yard. Since we bought the ranch, we don't have the time to go into town and golf. We used to golf every Wednesday night – well, that doesn't happen anymore!

The role of ranch women has changed since my grandmother's time. She talked about being out all the time, haying, driving the horses, checking the cows – she was like a hired man back then. And then work was harder, and as the years went by things changed – it was easier with machinery and equipment. My mom's still outside a lot, but I don't think it's as hard work as my grandma had to do. The changes have helped with the workload and the stress on everyone, but I don't think ranch women take as much time as they used to being with their family and spending time at home – they're so much busier now. I think there will be fewer women and fewer young families on the ranch, just because of the way the economy is. A lot of our younger friends who were raised on ranches, they're in the city and, for different reasons, don't want to come back to the farm.

We decide things together and set our goals together. Buying this ranch was our main goal and we achieved that. We'll set little goals from now on, and we'll work on that together.

I think if you really want to live on a ranch, I'd say, "Go for it! Enjoy the lifestyle, there's nothing else like it." It's a lot of work, but it's very rewarding and I wouldn't change anything. It's exactly where we wanted to be.

In the spring of 2004, Rajanne and Grant's son, Kolton, was born.

CARLEY COOPER WITH WAZ JEWEL AND WAZVAND STAR, 1964

CARLEY COOPER

I've spent my life with horses.

Carley Cooper ranches southwest of Tompkins, Saskatchewan on land settled by her grandfather in 1897. Carley was born there in 1925 and continues to ranch today. Her mother, Rose, was a great rider – she passed her skill and love of horses onto her daughter. Carley's granddaughter and great-granddaughter share this passion for horses. Carley started numerous Arab Horse Associations and was a founding member of what is now the Royal Red Canadian Arabian Horse Show.

THE RANCH STARTED with my grandfather, who was born in 1868. He came to Saskatchewan when he was eighteen years old. Then he went to Great Falls and while he was down there he married a widow with a three-year-old child. He and his wife had three children of their own – Rebecca, Elizabeth, and my mother, Rose. Not quite two years after my mother was born, his wife passed away with pneumonia. That August, 1897 he was ready to come to Canada, where his brother was, but the night before he was to leave, his horses were stolen, all but one, so he had to go out the next morning and get three more horses to put on his covered wagon. They came to Bear Creek, but then he took up this land right where we are today and nobody else has owned the land. It was his, and then my mother bought it in 1921, and then I took it over when we moved back out here in 1949, with Steve Schmidt as a partner.

My granddad worked on the railway when it was going through and the kids stayed by themselves. The oldest girl was nine. He was here at night, but he used to walk down to the railway at 6:00 in the morning, work 'til 6:00 at night and then walk back. He also trapped coyotes and they had wolves here then, and that is how Rebecca got poisoned and she died in '99 when she was seven years old. She is buried here at the coulee – of course we have our cemetery up on the hill. And my grandfather and my mother are buried there. My nephew was buried there, but when my brother died they took up the ashes and buried him in the cemetery in town.

My mother and her sisters went to a convent school in Fernie while my grandfather worked there because there weren't any schools near here. When Mother was old enough, she went to take her last schooling in the University of Alberta in Edmonton, so she had a better education than I got. She worked in a bank when she came back, in Tompkins. Later she went to Montana to her married sister's, and that's where she met my dad.

My dad was a very smart man. He had been a lawyer, but this was, I think, his third marriage. I have a half brother that was on my dad's side – he couldn't get along with my dad and he ran away from home when he was eighteen years old. He was working for the railway in Havre, Montana. He got his boot caught in the frog of the railway and the train killed him, and he is buried in Havre. My mother never knew him.

My mother had a tough life, but she grew up with an outward look of "life was good," even if you did have to work hard. I was born in '25, and I never heard my mother say a word against my dad 'til I was grown up. But he wasn't a helpmate, let's put it that way. My mother and my brother milked cows, and I think it was in '31, we lived off the cream checks, and he kept taking the cream checks and buying tobacco. And he wasn't milking any cows. It just came to the parting of the ways. He never supported us kids. My brother, Athlone, was seven years older than I am. So although being only two in a family, we always got along very good.

My mother was a great rider, and she passed that on to me. And she rode here. People came in and would build on the quarter or half section and she knew everybody who built and where they built. There were no fences, so they rode for the cattle and the horses, and she knew the country very well. She was very outgoing and she wrote about the horse races they had in Tompkins, and she rode in those races. My mother told me she was the first one around here to wear bloomers, and that was a disgrace! I have pictures of her and Aunt Liz and my granddad when they had been to a dance over at what we called the "Brick House," which was their cousin's place, and all the ranchers around would come there too. Sometimes we walked over. Pretty near all the time we went there for Christmas because there was music, dancing, food, and you knew everybody.

Growing up, I can remember coming into what they called the "Dirty Thirties," and there was no money. When we talk about no money, people nowadays don't believe that we sometimes didn't have money at all. You just got by on what you had and that was it. We used to get what we called "bum lambs" from the sheep ranchers. The ewe would have twin lambs and the ranchers would never keep their twin lambs – they always either gave them away or disposed of them. So my earliest memory is of my grandfather taking me with him in the buggy, very early in the morning. He would have to wake me up, but of course, if I didn't get to go I would cry. So we would go to Nicols' and Heffers' and to the different ranches and pick up any lambs that they would give away, and then we would have to feed them. It was on a beer bottle with a nipple, and you didn't do that just once a day, you fed them three or four times a day when they were small. So there was lots of work connected with it.

I remember when I started school. I was six years old and my mother took me to the school in Tompkins, two and a half miles away. Of course I shed a few tears – I had a very good teacher, Miss Abbie, and I guess I appreciate it more now because I don't think I would have made it through the first year without her. My mother left me at the school and I beat her home, and she had the horse! Of course she put me back on the horse and took me back to school. So that was my starting of school in Tompkins. I can't say that I enjoyed school when I was small because I used to have to ride to school with my brother. We had two gates to open and he'd jump off to open the gates and when he jumped back onto the horse, sometimes he'd knock me off. I used to have to step on his foot, and he'd take my hand and pull me back up and I'd sit behind him. When we got a little bit older, I got a horse of my own, and that was a lot nicer. If it was real bad weather we wouldn't go. When my brother was out of school and I was riding to school by myself, I had a horse that all I had to do, I could put my head down and just let her go and she'd bring me home. And I can remember lots of days that I was glad to do that because I was coming southwest, and the wind would be blowing lots of times from that way and you couldn't see, so I had a very good horse, which was lucky for me. The horses would be put in the barn when you got to school, before nine, and it was four o'clock then before they let you out of school. So we usually tried to have a little oats there, and we gave them a drink. Of course, boys being boys, they got trying to make these horses to buck – of course my horse had never bucked in her life, and they would tease her, and I guess I did have a temper and I can remember one time this boy was teasing her and I got so mad I just tied right into him. He could have beat me up if he'd of taken the notion. He never bothered my horse again, though!

I liked all animals, but I loved horses. It's brought me a lot of happiness because I've travelled with them. After I got my Arab horses, we started different shows and clubs and I showed my horses in Alberta, Saskatchewan, and BC. Then we went down to the States and showed horses down there. I used to be a very shy child, and by showing horses, it really helped me. I met lots of people. I still correspond with people I met, especially from the States. And it brought many friends, many trips and good times, so they've done a lot for me.

We've always had horses here. The first I can ever remember was purebred Clydes, and I got the job of cutting hay, and I wasn't very old. My grandfather was still alive. He died when I was twelve, so I imagine it was when I was about ten that he first took me out to drive the team on a mower and cut hay. Of course, he would be out there most of the time, not too far away, doing something, maybe hauling hay, but lots of time I was out there on my own, too. So horses have always been a part of my life.

In 1953 I started dealing on a Palomino stallion in Alberta. And I wanted him very badly because I had some light mares and I wanted some foals. Well, I didn't have the money to buy him. So I happened to be reading the *Western Producer* and I saw an Arabian stallion advertised at Lewistown, Montana. So I started to write – it was a doctor who owned him, and before we made any agreement the horse got kicked and his jaw broken, the lower part right by the chin, and both bones were broken. Dr. Simon wrote and told me they were going to put the horse down. We didn't have a phone in our house then, so I went to town and phoned him and I asked him if he would sell me the horse. He said I could come down and maybe we could come to an agreement. So my brother and my sister-in-law came with me. We had an International truck and we put the stock racks in the back, just laid them down. And I had never been anywhere near Lewistown, Montana, let me tell you. But I knew which way we had to go and I knew we had to cross the river. We crossed the Missouri River south of Big Sandy, and we went to Lewistown. We went out to look at that horse, and he was just skin and bones. And my brother said to my sister-in-law afterwards, "I know she's going to buy that horse." And I did. I bought him for $150 and the doctor had been asking $2,000 for him. And I had an agreement with him that if the horse was still alive in three months, I was to pay him another $150. Well, I had never come across the line with any livestock and I had never dealt with brand inspectors down there. We went to my uncle's ranch in the Bear Paw Mountains, and I knew I could stay there with him until I could get ahold of a brand inspector. You have to have an inspection from county to county down there. And I knew I had to have a veterinarian or a brand inspector meet me at the line. We phoned and got all the arrangements made. Then we came across the line with that horse and brought him home.

So that was my start in Arab Horses. I nursed that horse back, and I took seven pieces of chipped bone out of his jaw over the next month. He was ten years old then and I had him for ten years. He was well broke to ride. He was bred by the President's son, Elliot Roosevelt. His name was Sakib, and his American registered number was 2851, that's all there was in registered Arabian horses there then. And his number in Canada was 163. That was the second bunch of Arabian horses in Saskatchewan, was mine. I raised a lot of horses off him. In 1956 I travelled to Corniea, Nebraska and I bought a purebred mare. That was the start of our part-bred and purebred Arabs raised here at Hidden Valley.

The horses have taken me all over. I've been to Oklahoma City – we went there to view the reining – and we took the horses down to Albuquerque, New Mexico in 1991, and there was 1,600 Arabian horses there, that's the U.S. National Champion Show. I didn't ride my mare, Sharon McLean did, and they had two more of my breeding there and they came home with three top ten, and if you can imagine, showing against millionaires and coming out with three top ten, we were pretty proud people.

I have been involved in some ranch-related organizations. First, right after I got my Arab stallion, I got in touch with a couple who were at Vancouver and they had started the Arabian Horse of Western Canada. And I got really interested with them because I had this Arab stallion. And they encouraged me in what I wanted to do. I met Lenora and Stan Wilson from Calgary and we decided we were going to have a meeting and they were going to form an Alberta association. So, we met the first time in 1955, I think, at Chick and Lillian Miller's home in Olds, Alberta, and I think there was eight or ten of us. And they formed the Alberta Arabian Horse Association. And it is what's still running there in Calgary today. And then after we got the horse show started there, and I had bought this purebred mare, well, the first show I think was in '57 and I took her up there to show her. We put on a show every year, and we would have a championship class for mares and for stallions, and they would have to earn points in other shows to be able to show in those two classes. But they could come from anyplace, and see, there was a lot of horses came up from the States. But this was to be Canadian, these two classes, and that is what started at that time in Calgary and is now the Royal Red in Regina. And this is many years later, and that's a multi-million dollar business, that Canadian Arabian Horse Show. So from there, then I got together with friends here in Saskatchewan and helped start the three Saskatchewan associations and several light horse shows and light horse clubs. Anything to get people started with horses was right up my alley.

I still have Arabs. Diana, my granddaughter, and my great-granddaughter, Carley, just last year took two – the one mare that we did show all the time, Carley is riding her. And if you can imagine, turning a reining horse into a child's horse. Well, she's been riding since she could walk, but she was six when she took the horse up there. That's a great horse for her – she's so well trained and she never makes a wrong move. I'm not nervous with my great-granddaughter on her. It's just like I told my daughter when my granddaughter started riding, and my daughter is quite nervous with horses – 'course I got her bucked off a horse once, maybe that's why! But I used to tell her, "Stay in the house and don't look out the window." I took Diana, she was born in 1969, and I took her to Pat Wyse's riding clinic in Montana in '91, '93, and '95. She has a love of horses, too, and she's a natural born rider, which is good.

I did become involved in work other than ranching. My mother wasn't all that well and we got living in town for a while. And I worked in the café, as a waitress, but before that I was going to go back to school but I had to go out and earn some money for my clothes and my books. I had gone through grade nine, and if I can remember right, it was in the fall, and the war was on and they didn't have men for the threshing outfit. So I found myself with three other girls, working on a threshing outfit north of Gull Lake. And one of the girls was who my brother married afterwards. So we worked, we had a boss who had the threshing machine and we worked for him. Pitched sheaves, if you can imagine. Of course he was supposed to be pitching them too, but he was smarter than we were, and he would fill the outside of his rack and there'd be not as many in the middle and it would look like he had the rack full, and then he'd be after us because there'd be two of us pitching on our rack. And we didn't know enough to leave that space he left – we could never keep up to him. But, we'd get the best of him when we'd get into the threshing machine because we soon smartened up. With one on each side, when we wanted a rest, we'd throw the sheaves in there crossways and plug up the machine! But my brother kind of got even with him when he come to visit, 'cause he filled up the boss's rack and he filled up the middle too, so he had as many sheaves to pitch as we had.

In 1943 they were advertising for girls to go to Toronto to work in the small arms plant to make the guns, and they would pay your way down there if you wanted, or you could pay your own. Well, I was very lucky. My mother insisted that I pay my way, and that made a whole different thing when you were down there. Because the girls who allowed them to pay their way, they got the dirty end of the stick, and they were stuck there – I think it was for two years they had to work. But that was a whole new experience. I went down there on the train and thought I was never going to get there. Of course, we had been friends with people by the name of Hickman for years, and Betty and I were a few months apart, – we had grown up together. But Mrs. Hickman had moved down to Toronto with her family. Her husband had been killed in the First World War. And I went to stay with them while I was working, which was really good because seventeen years old, you can imagine how homesick you got! Mrs. Hickman used to have the soldiers come there, any of the boys from out here stopped there – it was just like a bit of home.

The first two weeks we worked in the plant we couldn't shut our hands. We were working with steel and they used an oil and water for a lubricant and the pieces of steel would get in your hands. And they were so swollen and so sore that you just couldn't close them. Of course, you toughened up after a while. But you sure felt sorry for yourself. I worked there in the summer of '43 and I came back just before Christmas, I think.

So that was a good experience for me. We made the Enfield rifle and the Bren gun in that factory, and I remember my first paycheque was $45 a week, before they took so much off of us for one thing and another. They came around and wanted us to buy war bonds – of course we did, and they just took it out of our cheques, so much a week. But out of that we had to pay our board and room and then we had a little bit left over. But of course, Betty being close to my age, her mother was real strict, but oh, we had lots of fun, we did lots of things that Betty's mother didn't know about!

Mrs. Hickman's oldest daughter was going with a fellow that had a riding stable. So we got going to the riding stable then, because we could both ride. I said I didn't think I'd ever pay money to ride a horse, but I went to that riding stable and paid $5 an hour to ride. And pretty soon they found out that we could ride, so they used to let us take the people out that came to go riding, so then we had lots of fun that way. Well, we met a couple of boys and I can remember this one boy, we used to call him the Yodeller – he could yodel so good – and his parents were, I think, from Switzerland. And this, of course, was quite good because we would get to see these guys. They'd come to go riding and Mrs. Hickman didn't know a thing about it. At least we didn't think she did. Or we would meet them to go to the show. We'd say we were going to the show, and they lived on what they called Brown's Line, which wasn't too far from where we worked. I have pictures of them. One time they come and called for us, and we weren't home – we told them not to come to the house, and she put the run on them. But after I came home Betty wrote to me and told me that the one that I really liked had been killed in the war.

But I came home, I really didn't like Toronto. It was cold and people weren't friendly, and besides, I like the prairies. So I came back and worked in town for a while. Then in 1945, in August, Christie was born and I worked in the café and mother looked after Chris. And I also worked for Mrs. Mussell, I used to iron or houseclean for Mrs. Mussell – they had the garage in town. And I never saw so many white shirts in my life. They changed, I guess, every day, and sometimes, I know – I counted once or twice – and there was twenty shirts in one week's washing that I had to iron for Mr. Mussell and the two boys. I like to iron, but that was no fun.

There had always been horses and cattle here, but in 1949 we came back to the coulee, with Steve Schmidt as a partner, and he worked off of the place until we got started again, because there was taxes owed against the place. And Mother did the cooking and looked after Christie and we did the work. And Steve worked in Mussell's garage and then he went to work on the CPR. And we milked cows in the morning before he went to work, and fed lambs, and I fed lambs at noon and again at night – that's what we did besides the other work. There was hardly any fences here anymore, so we had to rebuild ten miles of fence. I did the same kind of work that a man would do – cutting, baling, and stacking hay.

It really was difficult to be unmarried and expecting a baby. It isn't like it is today. So like I said, I always had the backing of my mother, and I think in ways – I hope – I paid it back because when I got showing horses, she travelled with me all the time. I don't think I did make many compromises after Christie was born. But I was fortunate because I had my mother, who supported me, and if I couldn't look after Christie, she did. The only compromise, I guess, is we didn't have a road into here when Christie was old enough to go to school. We just couldn't get a road from the Municipality and it was just like a prairie trail. And I can remember, Christie wasn't old enough to go to school then, but we had a terrible winter in '49/'50 and we couldn't even get down the coulee with a team of horses. So we used to walk to town, Steve and I. And if we had cream – we were selling cream again – we carried the three-gallon can. And I can remember one time we walked to town, got the groceries and the mail, and walked home again, and I let Steve talk me into walking over to Heffer's to the brick house, two and a half miles away – the snow was very deep – to visit and play cards. We went to come back, I suppose, ten o'clock at night and the snow was so deep, I thought we'd never get home. I never did that again!

I taught Christie until, I think it was grade five. Because there was just no way, in the wintertime, we could get to school. It was through correspondence school, so I had to take time out for that. But otherwise, you still had to do your work, you had your cattle and your horses to feed and things to do and wood to get. We used to saw wood. I can remember sawing wood, that was another thing we decided one winter to make extra money. Jim McEwan had a team of horses and a sleigh that was kind of like a box, and it would hold a cord of wood. And so we had one of those big circular saws, so we decided, here in the coulee where there were lots of trees, that we would get the trees out and saw them and sell them and split the money. So I would drive the team in the bush to haul the wood out, and when we got a pile big enough, Jim and Steve would quit cutting them down, and we would stop and saw them up. Maybe you think that wasn't fun!

I think it's a good life, a good place to raise children. I think you can make it a happy life – at least it's been a good life for me. Sure there's lots of things I've done without, but there's lots of good things I've had. And one of them, I think, is peace of mind. And you have to mix with people, you have to stay ahead, or stay even with what's happening in the world. But you also have to have a peace of mind yourself. And I'm fortunate, I have a very good relationship with my family. So I'm happy that way. I'm able to get around real well, and I drive and I'm able to jump in a vehicle and go where I want – I've always done that. And for me, I have good health, I think, it's not been as hard. But I feel sorry for the people, the older people who haven't that and have to go to the doctor a lot or go to the hospital. It must be terrifying for those people anymore. I hope I never have to move off the ranch.

George, my cousin, came in about 1983 and he made his home here. We became very good friends, like brother and sister. He died in '93, following an operation, and two weeks later Steve died. Two weeks apart. And my daughter wanted me to come to Bassano, but I said no. And the first while, yes, I worked, you know, to keep from getting lonely, but you live with it. And you go on. I could work then just as hard as I ever had. I fixed fence, I did this, I did the other, but I decided I wasn't going to feed the cattle because using the tractor and the truck and working alone, I didn't think that was a good idea. My daughter wasn't very happy with me for staying here, so I make a deal with my neighbours, Ryan and Cindy McGregor, that he would come and feed the cattle, and it's worked out very well, and he puts up the hay. But I have horses and I have the heifers in here, the replacement heifers, and I have cats and my dogs, so I had plenty of things I could do. You know, it's kept me outside, doing the things I like to do, and now in the afternoon, I find out I can't work as hard as I did, so I go into town and play cards. I go in to the senior citizens. So I'm mixing with people, which is good. Of course, I go up to my daughter's and stay, but basically, this is home.

I think you should be happy in what you're doing. If you're not happy in what you're doing, you can have all the money in the world, but it won't turn out right for you.

DELORES NOREEN DRIVING MULES ON ROAD MAINTENANCE IN THE 1920S

DELORES NOREEN

I was always curious. If there was something in the barn,
I had to see if I could ride it and usually I could.

Delores Noreen, now in her early eighties, has worked for the Slade Ranch at Tompkins, Saskatchewan for almost sixty years. Jack Mackie, of Maple Creek, who worked with Delores in the late 1940's, says, "She's a good hand." Robert and Anne Slade's son, Ken, describes his time with Delores as "growing up with a legend." Another son, Art, a well-known Canadian author, says, "Delores was an avid reader. The only thing I liked better than going through her collection of western books was the fudge she made." She taught Robert and his four sons to ride, and they respect her as a tough, loving woman. Robert attributes his "cow sense" to Delores and refers to her as his second mother.

WE RANCHED TWO MILES north of Tompkins. I was always with my dad and I rode a lot. We chased cattle for lots of cattle buyers and we brought them into the stockyards in Tompkins. Then they were shipped out by train. We didn't have big liner trucks back then. One time Shorty Mackie and I were moving eighty head of yearlings from out here at the Nicol Ranch. We didn't have no road allowance fences, not like today. They ran ahead of us. Well, Jack Dorfman had bought these cattle, and he rode up and said, "Don't chase them so fast." I said to him, "I don't see no damn halters on them." They slowed down on their own.

I have one sister. She was born in '28, the year I started school. Nowadays they've got to go across the schoolyard to pick up their kid. I rode to school by myself. Back then you had to buy your own books, and then you had to pay to write your departmental exams. That was eight dollars, a dollar a subject, and that was a lot of money back then. I had to sell a calf one time to pay – well maybe I wouldn't have had to but – I'm independent.

We always had a milk cow at home, all through the thirties. My mother got a cream cheque, not very big, but it all helped, and sold eggs, ten cents a dozen, made butter, sold butter. She baked bread. I don't remember having a loaf of bought bread. And my mother was always home when we got home from school. Nowadays I think kids are just hatched. The old mother hen looks after her chicks better.

My first horse was Maggie, a Welsh cross, and I raised and trained a lot of colts off Maggie – had her until she died. I had a saddle when I started school, but my Dad wouldn't let me use the stirrups, afraid I'd get hung up. I rode bareback lots, too. One way to get on bareback is put some oats on the ground, get your horse to eat, jump on his neck, give him a kick, and he lifts his head and you're on.

I was always curious. If there was something in the barn, I had to see if I could ride it and usually I could. I've been out in some bad storms. One thing, if you get lost, you get off your horse, turn it around three times, and get back on and leave your lines loose and let it go and he'll take you home. And he'll hit every gate, too. Once I rode a two-year-old to town, and it came up a terrible storm and I was sure she was lost. I was sure she was going in circles but I just left her alone. She stopped at the corral gate. I didn't know where in hell I was – didn't even recognize the place when I got there. Oh, I'd have been dead with my toes up.

I've worked for lots of different people. Most places I worked outside and inside, too, but I preferred outside, usually had to milk cows. I worked at Pillings and I worked with men quite a lot. I was expected to do the same things they were doing, and better, too. I never got no favours.

One Sunday morning years ago, I was bringing cows to the Ford Ranch. You had to run these cows into a corral and they had to stand so long, and then he'd come and A.I. them. The cows went into the dam to drink and I just went in with my horse – and it was quicksand. I thought I was a goner. First thing I remembered was to kick my feet out of the stirrups. I got off my horse – that was a colt I was riding there, too. She just kept jumping to get out and she finally got up on the bank and she was played out, laid right down. And I got out eventually, kinda wet. It was a funny thing. My glasses were still on but one spur was gone. How that spur got caught, I really don't know, caught in her mane or ... I know it was sure hooked solid some place. So it's still in there.

Jack Mackie told me how to cure a horse that wanted to run when you tried to get on. You just put a loop around one front foot and tie the rope up to the saddle horn. Yeah, that really works. I had an Appaloosa that was spoiled when I got him. He was about nine years old, I guess. You'd go to put your foot in the stirrup and he'd take off running. I just put the hobbles on him and he fell on his nose a few times. Then he quit. That horse worked good for me, not so good for other people, but later on he got cancer and that was the end of him.

I belonged to the Tompkins Light Horse Association and we competed in the quadrangle. You have barrels and it's like barrel racing, going around four different barrels in a pattern.

I have a trophy that I won in 1968, club champion in the ladies' quadrangle. Then there was pole bending. You have poles in a row and you have to weave around each one to the end, then come back through the maze again. You have to have a good reining horse to do it. You can't knock any poles down – you get penalties. This buckle I have I won for pole bending in 1967.

Then we had the tire race. You had so many tires and you had to race down with your horse and get a tire and crawl through it and get back on your horse. Once I got a tire and then two other girls jumped right on top of me. But I still had the tire. When we got all through I was surprised that my horse was still there. And we had the egg race. You held an egg in a spoon and you rode, and then there was the dollar bill race, bareback. You had to sit on your dollar bill then you'd race and if you lost the dollar you were out. You didn't want to have no bounce.

One time there was a wedding dance at North Fork. Shorty Mackie and I rode the mile and a half up to Thiel's and went with Lawrence. We took a team from there and oh, it turned cold and stormy. We had to go way around, couldn't cross Bone Creek because the snow was three or four feet deep. We got back to Thiel's and the coffee pot was froze solid on the stove. We waited until daylight so we could see what was goin' on, then rode home just down over the hogsback.

I married Duff in 1940 and worked in Swift Current that winter – got married and went to work the next day. Duff had worked for Slade's before in haying time and we both came back and worked there. When Slades bought the north ranch in the sand hills, they had purebred Hereford cattle, and we would move them back to what we called the big farm in the winter. That's where the feed was. And later I calved all the cows out north, too.

At the big farm there was a bad storm one winter. It blew the top off the cattle shed up by the stone barn. Jack Mackie was workin' for us and he went to town and he couldn't get back home. He got as far as the graveyard and his horse's nose started to bleed, so he had to turn around and go back. It was cold, really cold. We had cows in everything. I had one cow in the corral and I thought, that's a funny looking snow bank.

And that damn cow – there was a cow under there! I had the chicken coop full of cows, and the barn full of cows, and there was eleven head we couldn't get in. It was probably in January and the cows' faces were iced right over.

At calving time we didn't have vets then like they have now. We usually got the calves – block and tackle – that's what I used all the time. If they're not coming right, you've got to try and push the calf back and start over. Usually if you can get them pushed back and then turned, you're all right. You've got to be pretty strong to push them back because the cow pushes against you.

I remember Robert's sister, Elsie, staying with me at the north ranch, before the other kids were old enough to stay over. I'd put Elsie's hair up in rags and we'd ride the eight miles into town and when we got to the school barn, I'd take out the rags and comb out Elsie's curls. Then Elsie would take her dress out of the saddlebag and put it on before going down to the café and store.

When I had kids around, they minded. You damn right, they minded. I babysat the Slade kids and taught them their numbers before they went to school. We played rummy, twenty-one and sometimes we played aces-to-kings.

I didn't take no guff. If you're comin' with me, you mind, or else you stay home. I remember my niece, Connie – this was later years. She thought she could do whatever she liked. My dad was out there, too. I needed him to help me with a bull or some damn thing, to get it in. I told them they could sweep the kitchen. "No, we didn't dirty it. We're not sweepin' it." "Okay," I said. "Do you want to ride?" "Yeah." "Okay." Come back, they didn't have it done. There was this little girl along, so I gave her a ride and then I went and turned the horse out. Ho, ho, ho. "Well, I told you to sweep the kitchen. You didn't do it so you don't ride." They never pulled that stunt no more. That little girl, I had some chop on the truck – we had some weanling pigs, too. She said, "What can I do?" I said, "Well, you can sweep the truck out." Well, she hadn't started school yet and she had the truck all swept out, so I gave her a ride, you know, double with me.

Yeah, I grew a big garden and shelled peas with a wringer washing machine. I'd can and bake pies and take meals to the field. There were always cows to milk, too. And I taught Robert's wife, Anne, to make bread. She learned how to make good bread.

I had this kind of a stroke. I've got to take a bunch of pills now. I've got some arthritis in my legs, too, but I can get around pretty good. I was out at the ranch all summer.

I think ranching will always keep going. I really think it will. I don't see how it can quit. Like Brett Slade, he's a young person and he likes ranching.

Delores Noreen passed away in the fall of 2004.

HILDA KROHN

I don't care where you go, the bald-headed prairie – no matter where you look, there's beauty. It's all there if you want to see it.

Hilda Krohn (Baker) has had a lifelong love affair with horses. She has worked on ranches all over southwestern Saskatchewan. Leot Sanderson said that in the seventeen and a half years she worked for him, she never once booked off sick. She has a passion for the prairie landscape and the animals that inhabit it. In 1985 she and her husband, Lloyd, moved to their retirement home in Carmichael, Saskatchewan, where at age ninety-one, Hilda still resides.

I GUESS I WAS THE ONE that you dare me and I would do it. We were little kids playing in the straw pile. Bernie went over first, Kay went over second, Hilda came over third – about the fourth try, Hilda hit the top and she never went over. They looked all around the stack for me and didn't find me so they went to the house and said I'd disappeared in the straw pile. Daddy used a fork and Mother and Aunt Gertie used their hands, digging the straw away, and finally they drug me out. I was still alive but sputtering. From that day to this, I will not go up on a straw pile.

HILDA KROHN WITH HER DOG,
NELLIE, IN THE 1930S

There was an uncle over in England – he wanted pictures of us. Bernie, Kay, and I didn't know what he wanted pictures for, but he wanted to adopt one of the girls. Daddy got the letter back and, "Mother," he said, "he's picked the girl he wants." And of course that was a shock for poor Mother. So he lined us three kids up, "Well, you're going to be one missing." So Daddy read the letter to us and Kay beamed all over. She says, "I wish he'd pick me. I'd like to go." Then he said, "No, he's picked Hilda." There was no way I was going. I said, "No, I've got my mother and my father here and my brother and my sister and this is where I'm staying." Sure, I could have had the moon, sun, and the stars, and all the horses I wanted and everything. Back then we didn't have very much but we had each other and we had a roof over our head and we had beds to sleep in.

Our schooling was very scarce in early years. Then we went to Prelate – I never wrote an exam all the years I was out there because I was taken out of school, taken down in the kitchen to clean and polish. I could do that with my eyes shut. Then on weekends Sister was baking bread and she burnt the tops of the bread and I loved this burnt crust. So she cut them all off and put them in the pantry with a knife and a dish of butter and called me in after school. "Hilda, there's no light. We have to keep the door shut."

The only surgery I had was appendix. One night in the wee hours of the morning I was wakened with a terrible pain. Well, poor Mother gets up and she gets hot water bottles and covers me with hot water bottles. She didn't know and of course I didn't know. Well, it was getting worse so Daddy phoned Holdings because they had a coupe car. There was a snow bank – it was in March – outside the gate, so Mr. Holding got stuck in it. So Bernie hooked his saddle horse on – he was good with a rope and he pulled him out. Took me down to

Gull Lake, and the nurse took one look at me and left. To me, it seemed, "Oh, look at the angel floating down." She'd gone for the doctor and then I heard this man's voice say to Daddy, "She's got an hour." They told me afterwards my appendix had turned upside down and had gangrene. I was in the hospital ten days.

Years ago a fellow named Dick wanted horses broke in, so we'd get two horses at a time. I'd have one with my outfit and Bernie'd have one with his outfit, and we'd break them in. We'd use them for two weeks and he'd bring two more over and take the two broke horses home. Bernie and I were just little kids and we broke no end of horses.

Daddy had rented the land east of us and I was down there with six horses abreast on the harrows. We had no harrow cart, so I rode the horse behind and I finished harrowing and I came home. I couldn't cross the bridge down east of us with that outfit. The creek was dry, so I drove down in there. I never will know what scared those six horses. All six jumped into the lines. Of course, the horse I was riding didn't jump right away, so they pulled me out across his neck and I'm hanging there trying to stop them. I kicked him in the ribs. Well, he went up and I had to stop him because the harrows were all coming up and the harrow stretcher broke in the middle. So I dropped the lines – threw them away – let the six horses run. Right away I said to Socks, "We've got to stop them." So I rode in front, which I knew was stupid – six horses coming at me – and I said, "Socks, you get ready, 'cause they're going to hit you." Well, Socks must have understood because he lit into them as they hit him, all six of them, and they moved him a good three feet. So I got off and talked to him and patted him, and I got all my lines gathered up from here and there and drove into the yard. I put the horses in the barn for dinner. Dinner started very nicely, then Daddy jumped at me. "Well, what did you do that for?" So I excused myself from the table, walked out, unharnessed my horses and turned them all out. He watched me from the kitchen window. Later he apologized and I went out and brought all the horses back in and harnessed them, let them finish their dinner, and I went and finished mine and went back to work again.

Then I had this chance to go to Colonel Greenlay's northeast of Climax. They raised Thoroughbred horses and Hereford cattle. Colonel Greenlay kind of watched me and he was very particular about his horses, and he asked me one day, 'cause I was helping break these horses in, if I'd ride his horse in a horse show in Maple Creek. He rode one that he'd been riding for two or three years that he'd broke. She was well broke – a beautiful animal – and I was riding a three-year-old just broke in July and the show was in September. The judge calls me out – I'm on this three-year-old – "Miss Baker will you walk, trot, and canter, please?" So I walked, trotted, and cantered – turned around and came back – backed in and he placed me second. Then the judge asked both Colonel Greenlay and I to come out and go through the performance again and we did. He said, "Well, I guess it is Colonel Greenlay because his horse is a little bit better." Well, you know, I was as proud as a peacock that I'd made Colonel Greenley work a little bit harder just to keep his position of number one.

The three Greenlay sons treated me like a sister, and Colonel and Mrs. Greenlay treated me like their own daughter. I came home in '37, in September. I'd been gone for three years and Daddy and Mother were standing out on the road because they'd seen a rider coming from the east, and the little dog, Nellie, she was with Mother and Dad, and all of a sudden she took off and she came running down to the bridge to meet me – she'd never seen the horse before!

After I came home from Greenlay's, I ran the place until Mother and Dad passed away. Then Bernie sold everything and I had my daughter, Edna, then, and I worked in Bernie's

Café, which I didn't fancy, but anyway it was a job and a roof over my head and I was making money.

Bernie introduced me to Edith Wilhelm, and she asked me if I could cook. I looked at her and said, "Edith, I can't boil water without burning it." She said, "Are you willing to learn?" I said, "I sure am because I really need the job." "Well," she said, "that's all that matters."

I had my own little cook house. It was a bedroom, a kitchen, and a dining room. There were five of them. I thought, I can manage five. Well, springtime arrived and I ended up with sixteen. How am I going to manage this? Edith was very good. She made the pastry, all crumbled up, then I'd take out what I wanted and put liquid with it. Well, that went over good. She said, "Can you bake bread?" I says, "I never baked bread in my life." So she told me what to do. The first lot of bread rose about an inch. I said, "Edith, I'm going to feed the chickens." She said, "No, you're not." Anyway, those guys ate all the bread. Not one of them complained and the next batch of bread pretty near blew the oven apart. And I never had a failure of bread after that.

I took my saddle with me everywhere I went, and I rode around their hundred head of purebred Herefords twice a week. I never got tired of riding and I milked the cow. Well, milking the cow was quite a feat. You looked this way and that way because the place was crawling with rats. They ran around your feet – I'd reach in the chop bin to get chop for the cow and come up with a rat instead of a pail of chop. I never got bit, but I couldn't let Edna outside to play. One night this fellow said, "I'm going to put the big truck just east of the cook house, and when it's real dark I'm going to put on the lights. I hope everybody will be watching." "Oh, yes, I'm all eyes," The whole yard was just a moving mass of rats. But we were very lucky. Ferrets and weasels moved in and the rats moved out. Life went on.

I worked for Angus Willet on the Whitemud for a year. I was home this day and he had a Jersey milk cow, and it used to wander the yard. Edna went out to the barn and was playing with the calf. The cow was wandering around and she went up to Edna and could smell the calf on her, and that cow had these stupid little horns. I heard Edna screaming. I looked out – the cow had a foot on either side of her and a horn on either side of her and she was just a-working her. So I let the collie dog out that had fallen in love with Edna. The dog grabbed the cow by the tail and the cow was gone with the dog hanging on to it! She never came back to the house after that.

When I worked at Greenlay's in the thirties was the first time I met Lloyd. He worked for the White brothers. In the wintertime he'd shoot a deer and bring it up to Colonel Greenlay's. Later on I worked at Greenlay's again, and Lloyd and I got together again at a dance. We got married that summer.

When we worked at the Rosses,' I learned to drive a tractor. Never drove a tractor in my life and Lloyd came around the first two rounds with me. We had a plough with a press drill. "Now," he said, "you're on your own." So I cut down to the first corner. I was so used to driving the horse machinery and the gang plough, you know, you tripped it with your foot, so here – am I stupid? I'm swearing and finally I hollered, "Whoa!" But the stupid tractor didn't stop, and I looked back and Lloyd's almost falling off his tractor with laughter. You live and learn and, well – you find something comical no matter what you're doing.

We went down to White's and worked, and they had a twenty-five-year-old horse called Shorty. We'd pop Edna on top of him and he could cut out cows – she didn't have to. The only thing that bothered me, and it also bothered Lloyd – there was a bull there that would take a rider every time no matter who or what. And we were riding, bringing in the cattle, and Edna was with us at about five in the morning, and we'd got Dick Greenlay to come and help, 'cause you know – two grownups and a child with five hundred head of cattle is quite a bit to bring in. We got on top and the cattle were all at the east end, and I said, "Look, the bull is looking." Daddy said, "That's fine – I've got the persuaders here." So we gathered the cattle and the bull came to the back, and that was his mistake 'cause Lloyd could handle a bullwhip. And he just cut knotholes in that animal's behind like you wouldn't believe. Well, it wasn't long before the bull got to the front, so we were okay. Then when we were headed down towards the corrals, there was the river crossing and we knew when the cows got there they wouldn't go to their left to the corral. We didn't know where Edna had gone, so Lloyd took off and he got down there to the corrals and the river crossing and guess who's sitting on a horse in the river crossing? Edna. She knew the cows, when they came down, were going to cross and get out the other way. So she went down there to turn them in and Lloyd couldn't believe his eyes. So we got them all corralled.

We worked at Leot Sanderson's for seventeen and a half years. Leot and I did have our disagreements at times, over cattle, but that was very, very seldom. I loved working there. They were a wonderful couple. We were out every day. It didn't matter what the weather was. We'd feed the stuff on the north side – that's where we had our cows there and the ponies were there. Then we'd feed the herd bulls on the south meadow and load up and feed the heifers in the west field, and then we'd head west seven miles and load up over there and feed, and then back home. You feed where it's sheltered and you make sure the water is open, and you don't do all this in half an hour.

I bought this pinto, Patches, at a sale in Maple Creek. I rode her that fall out north gathering and there was a bull out there – he ran at her and hit her in the shoulder, so after that she wouldn't crowd on a cow, but she learned fast. So I rode her that winter and I went to get a cow, and coming down the hill the horse was doing something. I said to Lloyd, "What the heck is she doing?" "Well," he said, "all I can say is – her feet are not touching the ground." I said, "Thank you." I rode her after that and she was good until we were up on top one day, and I said, "What on earth's she doing?" "Well," he said, "I still can't tell you because her feet are never touching the ground." "Oh," I said, "thank you." So I said, "That summer fallow field, I'll take it out of her." So I ride over there and I get her into full gallop and I whip her all the way down the half mile and all the way back again and when I got down to the end – oh, she could work cattle like a dream. Well, it wasn't long after when she went wacky again. I wasn't chasing anything, so I asked the vets in Maple Creek to check her, and they said, "We can't see anything, but when the wind blows – is this when it happens?" I said, "Yeah." "Well," he said, "she cannot stand wind in her ears and when you ride where it's all enclosed, she'll go batty." So whether she'd been injured – who knows, but she turned out to be a beautiful, wonderful little horse.

I knew my little mare would take me home, 'cause I tried her. One day when I was riding down through the bush, I let Patches have her head and she took me to the exact spot where I'd come up. Then I knew I would not get lost. I don't think I ever covered all of that reserve in all the years I was riding there. It's a lot of bush. To me, it was a new excitement every year because I'd see different places each fall.

When Lloyd passed away – the funeral was on Friday – early Sunday morning everybody left, the whole issue. That night I could not go to bed, I could not sleep. I sat on the edge of the bed and I cried and I cried. Mindy, my little dog, was up on the bed trying to console me and I've never seen an animal like that – got her little head under my arm and those little brown eyes looking at me, then she had tears. Well, finally I collapsed on the bed and I woke up in the morning and she was right up beside me, and she sleeps, ever since Lloyd's gone, on his pillow every night, watching me. The minute I move, those little brown eyes are wide open.

I've still got my saddle and it's going to go up where Lloyd's saddle is, to Abe Nickel's son, Bruce – he wanted Lloyd's saddle. Bruce and my grandson, Bobby, got along like brothers. Bruce took a picture of it and sent it to me, and I didn't recognize the old saddle. He stitched it all up and built a saddle stand for it, and they have it in his front room, if you please. And that's where my saddle will go, where it will be looked after.

I turn the radio on first thing every morning. My sister, Kay, was here one Sunday. "Now," I said, "be quiet. Don't talk. Listen." And she couldn't believe what I listened to every Sunday. So I said, "I get my religion from all churches." And you know, I enjoy it.

I feel sorry for the kids of today because everything's all mechanized. They don't get to learn an animal from the ground up. I don't care where you go, the bald-headed prairie – no matter where you look, there's beauty. It's all there if you want to see it. I told my granddaughter, Cindy, "You listen to the trees up here at the top, the leaves. They're saying something, too, if you're listening."

GAYLE KOZROSKI

All my life I've thought about living on a ranch.

Gayle (Robson) Kozroski worked for lawyers for twenty-five years before moving to Kozroski Family Feeders in 1998. Their ranch is situated along Bone Creek in southwest Saskatchewan. Gayle was employed in Saskatoon and Regina, but looked forward to each opportunity to spend time in the country. Gayle is in her mid-fifties, she has five adult daughters. This is the third marriage for both Dennis and Gayle.

HERE ON THE RANCH I do everything. Basically, I keep the house and do the cooking, but I love animals and I love being outside, so I do gardening and yard work, I feed the cats and horses, and I check the pens. We have a feedlot, so somebody has to check and make sure everybody's healthy. I spend at least an hour every morning walking through the pens and talking to the calves and making sure everybody's walking right and nobody's got lumps or bad eyes or droopy ears. You know, you get to know the faces and the numbers, and there are some that have afflictions, so you look for them every day,

GAYLE KOZROSKI (FAR RIGHT) WITH BRANDING CREW, 2000

and "How are you today?" We have a good cattle dog, and she stays by my side all the time and makes sure they get up for me and chases them around a little bit so that I can see them moving. It's just a nice, quiet time. It's so different from the pressures of the city and always going by the clock. It seems like all my life I've rushed. Out here, you don't rush, you just do what has to be done.

I became involved in ranching when I was a kid – I loved horses. I grew up in a small town and there were kids that were bussed in from farms, and if I got a chance, I'd go out to their place and we'd ride horses and milk cows. My uncle was big in the Hereford industry and I used to go and spend time there. We'd brush the calves and ride them and do all sorts of things. They were a 4-H family, so we got into that. Then when I was sixteen, my mother, who of course loves horses too, bought horses, and after that we always kept a few horses close by, so I've always been a lot with animals.

There were surprises when I came to live on the ranch. You know, having grown up in town and living in a city, everybody figures that farm life is just so simple. It's just down to earth basics, you get up in the morning, you go out in the field and you seed, or you go out and look after the cows – when it's time, you sell them and when it's time, they calve. But it's not that simple – farming is a real intricate business. There are so many things to consider that I had never thought about – what to feed, when to feed, when to buy calves and when to sell, what's the market like and what do you feed them, what minerals do you give them, do you vaccinate? So many questions, so many things that are involved in raising animals and growing crops. I learned a lot.

All my life I've worked by the clock. I think the biggest adjustment I made on the ranch was when we first had our first meal together, I remember Dennis going out the door and saying, "We're going over here, and we'll be back about then, and have lunch ready." They came home and we sat down and had lunch, and we had coffee, and we had more coffee, and we sat around and chatted. I'm looking at the clock and looking at the clock and thinking, don't we have to get back to work? But, "Let's have another cup of coffee." And we chatted some more. And I thought it was a little ironic, because I remember Dennis often saying about lawyers, how they take long lunches and don't really work! Mostly, lunches are family times here and we sometimes spend two hours eating and discussing. We try to set our goals then.

It's the relaxed atmosphere here, too. The fact that you go by when it's light and when it's dark, by season, not by the clock. You get up when it's daylight and you quit working when it gets dark. So adjusting my clock was probably the biggest thing. But the other thing was that every day was Saturday. All my life, you know, you work Monday to Friday, then Saturday you do everything you can't do all week, and you kind of get going in circles. Well, I'm that way out here – because there is no set schedule, you just do what you can when you can. And often you don't get anything done, but you start a lot of things. It's a different way of life.

I think Dennis appreciates my work – I know that he appreciates the cooking. I don't think there's a meal goes by that they don't say, "This is good." They really enjoy the cooking and they've even kind of spread the word, because a lot of people come here and say, "I understand there's good food around here." Dennis's son, Gordon, ranches with us. He's out every day and we always have a big meal at noon, so lunch is not lunch, it's a meal.

As a couple we talk about decisions, and I think probably Dennis makes most of the decisions because he's done this all his life. I remember numbers well, so when they want to know how many calves are in that pen, or how many cows have calved, or how many cows went to that pasture, they ask me. But when it comes right down to when we sell or when we buy, it's pretty much Dennis's decision. And Gordon is fighting for his own rights – he's gone to a different breed of bull and he's raising his own calves and taken over Grandpa's calves, now.

I myself have five daughters and Dennis has a son and a daughter, but his daughter lives and works in town. We occasionally call on her for help, but she has really no interest in the ranch and its operation. Gordon is involved here because he is sharing in the ranch duties and he has half the cow herd and some of the equipment. Eventually he will take over the ranch. The rest of them are not involved in the decision-making. My youngest girl lived here, almost two years, and she did a lot of the ranch work and she did share in some of the decisions. She learned to run the equipment and learned a lot. While she was here, she took a correspondence course in animal health and now is working on a pork farm, farrowing pigs, and enjoying it.

I'm not involved in volunteer work at the present time, and this bothers me a little bit because I grew up in a small town, where my father was principal of the school and I was into everything. Then I married and moved to Saskatoon and didn't really get involved there, but in time, got back into the church and Kinettes. When my kids got older, I coached and was involved in the ball community, and working full-time, and raising kids. By the time I came here, I was pretty burnt out. My life was so controlled, I didn't have a minute to myself, and I just thought, "I deserve to have some time to myself, I deserve to think about me for a while." And I haven't got involved in the community, other than supporting what Dennis is involved in already – Bone Creek Co-op, he's president of that and has been for years. We have all the meetings here. And he belongs to the Saskatchewan Cattle Feeders Association and to the Livestock Inspection Board, and we also go to Canada Beef Export Federation meetings at least once a year. I go with him and take notes and listen and try to learn, but as far as the community – other than the odd shower, and we go to church occasionally – but we just are pretty busy here and I really hesitate to get into the community again.

The best part of being outside here is the freedom – the wind in the trees, the smell, the animals. It's just a great world out there. When it's cold and ugly out, it would just be nicer to stay inside and not bundle up and go out and check heifers at three in the morning, or run out and pull a heifer. But you know, it's all pretty rewarding, and it may be hard to get your clothes on and get out there in the first place, but once you're out there, I enjoy every minute of it.

The worst part of indoor work is the time that it takes and the fact that I think it has to be just so. I'm my own worst enemy – I think things have to be clean and the laundry has to be done and the ironing, and the cooking, and it isn't just thrown together. You have to have salads, and you have to have meat and potatoes and vegetables and a dessert. So it's just more work than I really want because I'd just as soon be outside.

Health problems have affected my ability to be as involved as I would like to be. I came out here in October, and at the end of November, Dennis and I were in a car accident and I had a broken neck and I bashed up my ankle really badly, which turned out to be the major problem. The neck healed. I spent eight weeks in a traction halo and several months not being able to walk on my foot, but once they got all put together, I can walk, with a limp, but it really slows me down. It was the end of March before I got back on a horse. Everything is more difficult because I can't walk as well as I did. But I do everything. SGI tells me that I can do eighty percent, so I don't get anything from them. Simple things, like walking on rough ground, you hit a high spot and "Whoa!" It just takes longer to do the things that I want to do. I wanted so desperately to go out, there were so many things I wanted to do, and I couldn't. When I was in that stupid halo and I had a cast on my foot, I walked with a walker, so just to go downstairs to the cold room to get something was a major problem. Of course, I felt that I had to do everything that I used to do, whether it was difficult or not. I would try to get organized before Dennis went out in the morning, and I would say I needed this from downstairs and even a couple of things from the freezer out in the porch, which is a step down. Eventually, I got so that I would go downstairs on my bum and I'd gather things, and I'd just move them a step ahead. Even now, going up and down stairs, I stop and think a while, and I gather things up that I have to take down, and while I'm down there, bring all the things up that I have to bring up. Dennis had a bit of whiplash, as well. We took physical therapy in Swift Current, fifty-five miles away, which wasn't bad, but we were going three or four times a week, sometimes every day. We had to make a lot of trips back and forth to Saskatoon as well, because that's where my doctor was who looked after my neck and my ankle. So for at least six months we were running back and forth. We hired someone to come in and help, and Gordon pretty well took over.

I think you enjoy going out as couples, simply because in the farm community you tend to visit more because you're not with people all day, so in the evening, if you want to see people, you go out and visit. As for leisure activities, we take in dances or suppers, dinner theatres, weddings, but more than that, it's visiting with couples Dennis has been friends with for years. We talk about everything, lots of cow stuff! When I get together with just women, we talk about grandchildren and children and husbands, life in general.

When I was growing up, we certainly discussed topics like abortion because I, in fact, had a child before marriage and gave her up for adoption. But abortion wasn't really an option back then, it wasn't really considered, there was always another way. But then, I have five daughters and I know that my daughters, after the fact, discussed that several of their friends had had abortions, and I was quite shocked. But it was pretty much a common thing in their generation.

In the future I think women will have a lot more in the decision-making and control over the financing. Ranching is a big business – it needs good management. I think women may be a little more capable of handling the bookwork and the management side of it. Men seem to get carried away with the actual activities and don't have time for bookwork. They don't like bookwork. The other thing is that men are so afraid of computers, where women, I think, adapt more easily. I certainly do because I've worked with them in the office. Although young people, like Gordon are getting better – they come by it naturally, but Dennis won't touch them at all. He really relies on me to do the bookwork and handle the bill payments. If he wants to know how much money we have, or how much we don't have, he comes to me.

I suppose we will make a transition from greater to lesser involvement in the ranch as we grow older. I hope not for a while, but I imagine it will happen. We talk often about retiring, but I can't imagine going back to the city – Dennis doesn't want to – or moving into town and being on coffee row. I think, probably, if Gordon and Clare and the girls ever take over here, that we would perhaps build a house across the road or something. I don't know if that's possible, to back off and let him take over. But it probably is, I hope it is.

I certainly understand why people fight for the privilege to remain on the ranch. I would love to be able to help persuade the government and anybody who isn't familiar with ranch life that it is a way of life that we must preserve. But mostly I'm just grateful for the opportunity to be here.

CLARE KOZROSKI

I've had the benefit of being able to say, 'Yes, this is where I'd like to have my children grow up, these are the values that I would like to share with them.'

Dr. Clare (Cormier) Kozroski is a family physician who left a thriving practice in Thunder Bay, Ontario, and moved to the prairies when she married her husband, Gordon, in 1997. They ranch with Kozroski Family Feeders, southwest of Gull Lake, Saskatchewan. They have two daughters, four-year-old Gina and two-year-old Leah. Clare and Gordon live in Gull Lake, where Clare practices medicine.

I'M THE ONLY DOCTOR in Gull Lake, so I spend a lot of my time doing my medical work and little of my time doing immediate ranch work. But certainly, I'm involved in an extended ranch family. Gordon's a fourth-generation rancher – his land is not going to move very far, but it also means that his dad and his grandfather are still active on the ranch, his dad living there full time, with his wife, and there really being only room for one family in a modern ranch house there. So by virtue of my having to be on call twenty-four hours a day at least one week out of two, we stay in town and my husband

DR. CLARE KOZROSKI DRYING A NEWBORN CALF, 2000

drives the twenty-five minutes to the ranch. He goes there pretty much every day. Maybe one out of three weekends he'll take Sunday off to stay in town with the family, or we get a chance to go out to the ranch from time to time.

When we were courting, it wasn't made clear to me that there was really no viable option for him other than to ranch full-time. I knew that it was his dream to return to his home, to play hockey, to have children, and to be involved in some aspect of agriculture. But it never, until we actually moved here, became apparent to me that it would be so many hours per day, so many days of the year. In fact, I think he had sold me, a little bit, on the flexibility of his hours as a rancher, saying that he could be quite a good "Mr. Mom" in the house as well because he wouldn't have to toe the line for a boss other than his family. I don't know if he deceived me purposely! I think I was just naïve in some ways and knew that he would be a good dad and would be motivated to spend time with his kids, as well as time on the ranch, which to me didn't sound like "work."

We're pretty good at making decisions together. We're both good compromisers. Many of the ranch decisions, of course, are made between Gordon and his father at this stage. We had tried to have regular meetings, supper gatherings, for my husband, myself, his dad and his wife, to talk about how things were going with the ranching, so that we would all be in touch. That didn't work well because we didn't guard the time required. Financial decisions are shared to some degree, although I'm not aware of the day-to-day negotiations. By virtue of the fact that I have a job that the bank respects, I can sometimes help out with debts and renegotiations, repayments, and that kind of thing. I have nothing to do with money, but I do have a lot to do with decision-making, with how the ranch is going to work. Decision-making is also influenced by how I have Gordon help me in my work. In spite of his long hours at the ranch, he is my office manager and does all my computer work and trouble-shooting for my receptionist and the physical maintenance of the office and yard.

In our domestic work, I rely on Gordon a lot. We have a live-in caregiver. We built her a suite in the basement of our home in town, which is connected to my clinic, because, like a ranch, the business and the home are so close they can't be separated. There's one doorway between where I see my patients and where I see my children and cook my meals. Since I've been back at work full-time, our caregiver comes to work first thing in the morning. She has a long day and is ready to leave at six o'clock. I'm very often not finished my work at six o'clock, so I'm calling Gordon to remind him that this week I'm on call, or today in particular, I can't get away and would he please do his best to show up shortly after six so we're not abusing our caregiver or burning her out. Burnout is a big concern of all parties here. Our live-in caregiver, myself in my work, and Gordon in his split life as well. Neither he nor I are good housekeepers, so we generally just don't do that. The home maintenance is minimal and we make that choice that we won't live in "house beautiful" but we'll have more time with our children and put our priorities elsewhere. He's just as quick to do dishes or tidy as I am, and just as quick to bathe or feed the children.

The most common thing Gordon says about my work, at least to me, is that he's proud of me and he knows I work hard. He thinks I work too hard and he thinks we ought to get some more medical help in town so that the twenty-four-hour call is less draining and the administrative burden of my work is less. He sure doesn't resent my helping people – he does resent the time that's required to manage the office and the paperwork.

I have no fixed duties on the ranch because I'm unreliable in the performance of these tasks. I can't commit to do very much at all because even when I'm not officially on call, I'm often required to come back into town. So, I try to take time off, deliberate time off, during calving. I've succeeded in that for two years, at least, where I've spent the better part of a week full-time at the ranch, including sleeping there most nights during calving, and enjoying that exceptionally. Riding with Gordon, maybe one or two hours a day, helping him to tag, and doing those sorts of things. The other times of the year are sort of catch-as-catch-can. If we're out there I'll do whatever people accept my help in doing. I'm certainly not shy to get into the poop or any other perhaps distasteful or unsophisticated work, in some people's minds. I've been lucky enough to ride safely, so far, and help, of course, with veterinarian-type tasks, though I'm not a skilled veterinarian at all.

I first learned to drive in the Army Reserve. The first vehicle that I ever drove was a two-and-a-half-ton truck, which had an automatic transmission, so it wasn't all that hard. But certainly it gave me the feel of handling heavier machinery and not shying away from that, even though I'm not a large or powerful woman. I've also learned how to dig – to handle a shovel and get dirty and tired, or live in the outdoors, day or night. Those are some of the really valuable things that life in the field, in the Canadian Forces or any other forces, can teach you.

Once I got married, at the age of thirty-seven, I didn't know if I would be physically able to have children, so was still optimistic and certainly my husband was optimistic about having children, and we were glad that at thirty-eight and forty years of age we had these two. Contrary to what some women may experience, my pregnancy allowed me to take time away from my practice, and leisure time for me may translate to time free to go to the ranch. I did actually ride, pregnant, until maybe halfway through my pregnancy, at least the first one, having been lucky enough to be in excellent physical shape and not finding that a risk-taking activity. Having some maternity leave with both the children, six months with the first and four months with the second, that was the most time I had ever been able to spend at the ranch. Unfortunately, actually after the birth of the first child, I didn't recognize the help available at the ranch, and I ought to have, because I was suddenly isolated in town with the baby. It was the first time that I had ever been required to stay at home in my entire life. That was a good set-up for feeling somewhat depressed, with the hormonal fluxes and risks of postpartum depression, anyway. If I had been smart, I would have hauled myself and the baby off to the ranch and had that family support and warmth there and more fresh air!

Many of my best friends are women living on ranches, partly because the socializing that I do is limited and is very often with friends, our neighbours in the ranching community, with other folks in cattle organizations or co-operatives. I find it enjoyable to talk with them, and of course, I'm still learning great quantities of information and skills from those women. Sometimes I think that they are more likely to talk to me on a level that is unrelated to my medical work as a doctor, whereas the folks in town see me more often as a doctor. But I have friends there with whom I can curl, for example, or with whom I associate in church activities or parents' activities. My life is split between the ranch community and the town community.

I think back to Gordon's grandmother's activities on the ranch – her raising her five children, tending chickens and the milk cow, and doing all the wash by hand, carrying water and having no electricity for much of her time there, for example – things like that that we take for granted. Certainly, ranch women now have less demanding work in those ways because the ranch work is being automated, and the domestic work has been as well. People continue to point out to me that they had no phone to ring, no fax to bother them, they weren't expected to attend so many social committee or service club meetings, or do the services that Gordon's grandmother does now that she lives in town. Ranch women today are much more split into three rather than two, the two being mother and farm/ranch woman. Now the woman is more likely the one to have the off-farm income that is required to maintain a family financially. That's difficult. It's also difficult because they no longer have the benefits of isolation, yet still some of the drawbacks of isolation. If they do live on the ranch, because even if it just means commuting to that job that's bringing in some more money, that's another vehicle, and more time away. It's also that guilty feeling that you're never able to give enough to your job or to the family. Those stresses, I think, have increased over the years.

Ranch involvement has a huge effect on our leisure time. Recently it was Valentine's Day and we were talking about our big date for that night. It was the annual Saskatchewan Stock Growers' meeting, which turned out fine because it was a dinner and a dance and it was fun. We socialized with other folks who have similar interests to ours, in terms of ranching. So, I think, again as most parents of young children find, their adult social activities are curtailed for a while and then may recover again. But if our social activities, through ranch neighbours and friends, continues with dinners at each other's homes and such activities, I think that that's a very healthy social activity and one that we do share with our children.

The kids have been coming to the ranch since they were in the womb, not often enough though. They seldom go to the ranch with Gordon themselves because they're at the stage where he doesn't believe he can adequately supervise them and get anything done safely. So they generally go out to the ranch when I can go with them. Sometimes the nanny is able to go out with Gordon and the children. I think, as the kids are safer to have around the ranch activities, Gordon will take them out more often. Unfortunately, they're soon going to be school age and their time will be restricted to weekends and holidays.

Since they were very little kids, they've been near the livestock, especially. They absolutely love playing with the barn cats and seeing the deer and the antelope and the pheasants, and certainly, being on the back of a horse – not truly horseback riding. The benefits of ranch life are many. Some, from my medical/scientific point of view, are that children who grow up in cities or environments that are too clean or sanitized tend to have poor immune systems. They have more asthma, they have more allergies and other problems like that, that we've created for ourselves by getting away from the land and the animals and away from a more natural, down-to-earth way of life.

We both feel that you can become a very strong individual growing up on a ranch, finding challenges that are healthy for you, although some of them are risky. If you're going to take your dirt bike over every rock and hill on the ranch, you're going to fall every once in a while, but you'll learn how fast you can go, how hard you can go, how tired you can get, and the limits of your machine. We'd rather have our kids doing that, or riding the horses hard, than inventing silly challenges for themselves or with other youth who may be misguided. A big reason why I'm here is because I'd like my children to grow up on a ranch.

I think it does make a difference whether they're male or female, which is strange to some degree, because I've generally grown up with male colleagues and male friends. Several jobs I've tried have been in male-dominated fields – that's the stereotypical term for them – but I never felt that I was subordinate to any of my male colleagues. Now, I think that a woman should be allowed to do any kind of work that she is fit to do, mentally, physically, spiritually. I believe that women are physically not as capable at a lot of tasks as men are, and they should not be placed in positions where they cannot pull their weight, or pull the weight that a man could pull. But I also believe that there are different ways of achieving different tasks so that if a man might force something, a woman might "finesse" it or use another tool, or help the work in a team effort that could save time and energy on all sides, rather than doing it solo. I think a female could do just as well at some ranch work as some men, but I still don't believe – now I may be burned at the stake for saying this – but there are some parts of ranch work that I believe most men would be more capable of than many women. Now, my husband has said, prior to the time that it mattered what gender our babies were, that they would be doing everything at the ranch that any son of his would. Certainly I don't think the girls are going to be limited in any way. I think our girls would be allowed to do as much as they can safely do, and they should grow into more duties as well as fun and seasonal work at the ranch. And if one of our daughters, or both of them, wanted to ranch for a living, that would be quite all right with us. We'd be proud to have them as the female heads of that ranch and be the fifth generation of the Kozroski ranchers. Maybe women can overcome some physical inequality with other methods of getting things accomplished.

My philosophy of life has come through because I've chosen to become a part of a ranch family. I want my children to know what dirt is, I want them to know where milk comes from, I want them to know about life and death in a very natural way, as it occurs with plants or animals or people, and understand what this earth that we live on is all about. Not be isolated from that in a controlled environment.

Hopefully, in the future, our goal, in fact, Gordon's goal certainly, is to have us live at the ranch and still have me be able to practice medicine but have the girls grow up in more of a full-time ranch family. Living there and doing their daily chores and having that be part of their fibre rather than more of a visit, as it is at this stage. Rather than having this lifestyle inflicted upon me or being born to it, I've chosen it, and I've chosen it because I think that ranch life values and my life, perhaps not as a ranch woman at this stage, but certainly the wife of a rancher and a member of a ranch family, fits with what I think is valuable in life and what I want to pass on to my offspring.

In 2003, Clare and Gordon's son, Alexander, was born.

WOOD MOUNTAIN UPLANDS

For more than a hundred and twenty years, ranchers have raised livestock in the Wood Mountain Uplands, a hilly region that stretches from east to west for a hundred miles across the deep south of central Saskatchewan from the Big Muddy to the Frenchman River. The Uplands appear to divide the watershed. Streams along the southern slopes flow into the Frenchman River, Rock Creek, the Poplar River, and Beaver Creek. Most streams on the northern slopes flow into the Wood River, then into Old Wives Lake, and then underground through the Big Muddy Valley to the Missouri watershed. Although many of the coulees in the Uplands are wooded with aspen, Manitoba maple, willow, and a variety of berry bushes, the dominant plant species is grass. June grass, western wheatgrass, grama grass, and more than fifty other species flower as spring moves into summer. Just as Major Walsh predicted in 1878, Wood Mountain was as fine an area for ranching as would be found in the West. In an area where the population is estimated at one person per four sections, there is ample range for raising livestock.

The first ranchers were a motley bunch, including ex-members of the North-West Mounted Police, gentry from a foreign land, and politicians. To begin with, all of them were men, often bachelors. Some of them married Lakota women, women who knew how to live off the land. Others brought wives from eastern Canada, and those women usually fulfilled traditional roles of homemaking and mothering with little outside involvement in the ranching activities. Just after the twentieth century began, many ranchers arrived from the United States to settle in the Uplands. Their wives were often well acquainted with living on the frontier. They wore bloomers rather than dresses, they could ride and rope, and they could shoot a gun if need be.

Ranch women at Wood Mountain today wear denim jeans, usually an American brand, but they can still ride and rope and shoot a gun if they need to. Most are probably more involved outdoors than indoors. In the mid-twentieth century, one generation of women made the transition from an almost totally dependent, domestic role to one of independent involvement. Now young ranch women are often partners in every aspect of ranching, and they are raising their daughters to fulfil similar roles. The ranching future in the Uplands looks promising.

MARJORIE LINTHICUM

Cliff Andersons sister

I am as much a part of the land
as the coyotes and the gophers.

Marjorie Anderson was born in a blizzard on the LA Ranch south of Fir Mountain, Saskatchewan in 1936. She was the twelfth child in a family of thirteen. In Saskatchewan, she was one of the first women to become a full-fledged partner in a large ranching operation in the 1950s. Marjorie married rancher Frank Linthicum in 1960. In her unassuming way, she taught their five children all she could about living on a ranch. Marjorie died of cancer in March 2003.

In 1976 Marjorie Linthicum presented a brief at the hearings for the Grasslands National Park. In that brief she wrote,

I appreciate and respect first flowers in the spring, the wide starlit sky and the night winds whistling, the lonely call of the mourning dove and the thread of the long-billed plovers, the howl of coyotes, the waving throats of the sage hens, the smoothness of an antelope as it runs a short distance to pause and look back and snort its familiar warning, the screech of the hawks as I come upon a nest of their

MARJORIE LINTHICUM COUNTING CALVES
AT BRANDING, 2001

young, the vastness of the prairie under a blanket of snow, the threat of an approaching snow storm and the shelter of a brush coulee, the sharp thorns of the buffalo berry bush, the skill of red ants as they pack away crumbs from a noon lunch to some small stone. I know what it's like to ride all day and never encounter another person, to have a faithful horse bring me sixteen miles home through a blinding storm, to drive cattle home in the fall and have them strung out for two or three miles heading for winter pastures, to see bands of wild horses trailed out to packing plants, to sit on a knoll and watch cattle graze on an alkali flat or two mighty bulls battle over a harem of cows, to repair fence all day and pick wood ticks off all evening, to sit on the high butte and look south over the prairie for miles and miles, to drive cows and calves to summer pastures and then sit and watch until they mother up, to dream as a young girl of riding south to the badlands with my dad and then having this dream come true, to trail carloads of grass-fed beef thirty miles to the railway in thirty-five degree below weather, to see buffalo horns on the prairie and wonder if the buffalo died from a winter storm or an Indian arrow, to have a horse get loose and leave me fifteen miles from home. I am as much a part of this land as the coyotes and the gophers.

WHEN I FIRST STARTED to school, I rode with three other members of the family and when they got into high school, it ended up to be my younger sister and me and we had lots of good times riding back and forth to school. I went to country schools, two different country schools [Macworth and Sister Butte], and in the winter, because I had older brothers in high school, my mother would move to town. We didn't really have holidays, we just went to school all year round. In town ranch kids were kind of belittled. I think some of the kids thought that our parents were land barons or something, and of course kids pick up what they hear. And then we didn't live in a school district, so every place we went we were kind of outsiders at the school. That was mostly in the beginning and in the war years.

There was a shortage of teachers, so when I finished high school in 1955 I went out and did babysitting at a school in the country, only four kids but it wasn't what I wanted to do. I guess if I had wanted more education after high school I could have had it. Two of my brothers went to university and my four sisters were teachers.

I became a ranch partner as I got older – I partnered up with my brothers. My dad decided he wanted to let the ranch out on shares and I was to be part of it. He probably had a major role in that. I had been riding at calving time, riding, sorting cattle, moving cattle, helping with fence, and I would also do a bit of farming because there was farming to be done, too. Harvesting. Putting up feed, grain, green feed. Grain bales, no hay, we didn't have any hay land – it was all cropland that had been seeded for feed. I was driving a tractor on a binder and later swathing and baling. Hauling bales.

My mother did the indoor work, although I helped her with washing and some cooking too. We usually had hired men and my mother might be gone with my younger sister and I would do the cooking in the wintertime. I learned how to can meat, I helped her make soap a few times.

Not too many women had been a partner in a ranch like we had. Same with the farming aspect – not too many women had done that and it created problems because it had never been done. Women weren't supposed to be able to do it. So I kind of had to prove that I could do it before I would be recognized. I think I have had better opportunities because I was a girl. My dad was getting older and I was more his pace, so he could take me with him and it was easier for him to set the pace with me.

I was married in 1960. My husband, Frank, was in partnership with his father and brother, but I had land of my own and some cattle of my own, so I just took them with me when I got married. I probably went along with a lot of the same decisions the men made for their place. If they kept their yearlings I kept mine, if they sold calves, I sold too.

If I had extra grain, I could sell it but it had to go through the Canadian Wheat Board and the fellow from the Wheat Board said that a woman couldn't have a quota book, and I couldn't sell grain without a quota book. So the elevator agent came out and brought this fellow from the Wheat Board and proved to him that I could run the tractor. The CWB is really not fair but I don't feel that I really should have a big role in what the CWB does because the only grain that I sell is what isn't needed for cattle feed, and I consider myself more a rancher than a farmer. I just like cattle better, I guess. I make about ninety-five percent of my living from livestock.

After I was married, I probably couldn't do as much as I was doing before because I was looking after my family. I would take them with me, and then when they got older they learned to run the machinery, too. It's different now. Now my husband's brother quit and his dad is gone, and our oldest son is a partner with us. He's taken a lot of it over. Some of it has been turned over to one of our daughters. But we still are part of a lot of the major decisions.

My first obligation was to my family, and then from there I would go and help get the ranch work done. Sometimes schoolgirls would come and work in the summer. Probably a lot of that hired girl bit was reflecting back to my mother because she was always obligated to look after the hired men and I didn't want to be tied to a hired man, to cook for them and wash for them, and so I found it easier to have a hired girl and go and do what that hired man did than to have a hired man under my feet all the time. My mother probably never had the opportunity or she might have been as independent as I am. Of course, now it's different, everything is shared. At that time I don't think it was. My dad made all the decisions.

I've been involved with the Hereford Association, Canadian Western Agribition. I guess that one of the highlights was being one of the first women judges at Agribition at the commercial cattle show. I guess they thought I had a good opinion of cattle. And I was a member of the Saskatchewan Stock Growers. The women had their own programs but I didn't always go to them. I'd go to the meetings. I was more interested in what was going on at the meetings. In my mother's time, it was just an outing.

We've made a lot of changes, like utilizing a feedlot. Instead of growing the cattle out at home on the range, they're sent to the feedlot and finished and marketed, and I think we save marketing costs and save a lot of the responsibility of health concerns for the cattle because we have professional people looking after them. Probably we'll develop the kind of cattle that work best in the feedlot and turn them over quicker, where if we kept them at home to grow them out it takes more time.

I think the ranching industry itself has probably made changes before others even had a chance to say they needed to make changes. The ranch associations and stock growers associations and groups like that have the interests of the livestock, and I think they're taking care of things before the animal rights people tell them what needs to be done. I suppose there are situations where it doesn't happen, but most ranchers look after their cattle, so the cattle will look after them. In the last few years there's been much improvement in the ways that people are looking after their ranches and cattle and their land – I think they're taking care of it without somebody telling them what to do. They can see the advantages of doing things differently.

And the women are partners right in there with the men. Like two of my daughters have gone to agriculture college, and that might have changed how they operate their land, utilize different methods. There have been changes but I wouldn't say big changes. The bigger change was from the generation before me to mine. Like when ranch women started doing more ranch work and it was accepted that they could do it. In the future they'll be partners. It's all for the better.

I think there will always be ranching here. I think that's what this country is suited for, and as long as we're producing a product that people want, it will always be here.

LOUISE POPESCUL PUSHING CALVES
TO A BRANDING TABLE

LOUISE POPESCUL

Trust in the Lord with all your heart.

Louise (Linthicum) Popescul credits her parents, Frank and Marjorie Linthicum, for the ranching skills she learned while growing up on a ranch south of Glentworth, Saskatchewan in the 1960s. And yet she sees herself in a somewhat different role, more of a partner on the ranch where she and her husband, Wade, raise their three children, Jesse, Tyler, and Jade. Louise carefully considers the changes that will be necessary to ranch in the twenty-first century.

DAD WAS IN CHARGE of the ranch. He was the one who made the final decisions, but I know Mom had lots of influence. They kind of balanced each other out. Mom was the conservative one and Dad would be leaning more the other way, and in the end they met in the middle. I always expected to be working outside, more like Mom, I guess. We always had someone living with us in the summer – hired girls to look after us while she worked outside. I swore when I had my own kids I wouldn't have anyone living in my house. No hired men and no girls either. To be honest, I don't know why I felt that way – probably because we hardly ever saw our parents in the summer time. They were always outside working.

I can remember being in grade twelve and not knowing what I was going to do, and I had all kinds of ideas – most of them involved going to Alberta. Mom thought it would be great if I did something with animals, something I could use on the ranch, that sort of thing. I studied Animal Health Technology at Olds College. I worked at some feedlots, vet clinics, and for an embryo transplant company near Saskatoon. Between years at college I even went custom harvesting.

I guess when I worked at the feedlot, the first feedlot I worked at after I got out of Olds, I was given everything to do like the rest of the workers, working with cattle, and they also had a big farming operation and I did that stuff, too. I think some of the men felt threatened a bit, just by the comments I heard. They probably weren't used to working with a woman at all. I know Mom used to tell me what it was like to be a woman back when she started farming, and as we grew up she always had her own cattle and her own brand and her own farmland. But I never grew up in that part of the century when women couldn't have that stuff. So I didn't have the same feeling toward those things as she did because I swore when I got married there was not going to be two branding irons in the fire, and everything was going to be ours and not mine and his, just to make life easier.

When I came home I wasn't married yet. That's when Murray, Melody and I bought this place, it just sort of fit into Mom and Dad's ranch. Then after I got married, a few years later, Murray and Wade and I lived here together, and I always had a job at the stockyards and some of those places, whatever. Then before Wade and I had our first child, Murray and I talked and we split everything up, and Murray got into Dad and Mom's ranch then. It was time, with a family coming, to make that decision because we just kind of lived here and worked here and did our own things on the side. When we had kids, I didn't do as much outside as I did before. Wade handles most of the stuff outside and I kind of do the inside, and drive the kids around, work wherever I'm needed. I help him outside, but try to let him make the final decisions, I have input, but in the end it's his responsibility to look after us. I have no doubt that if I had to do it, I could. It would just happen.

Priorities. I've been trying to figure that out for a long time. Some people have dreams or passions or whatever, and I really don't, not that I can think of. I don't know, and it's strange, I know that the ranch was Mom's passion because that's what she lived for. I just want my kids to grow up and be decent, honest people, to respect other people and maintain a relationship with Jesus. Relationships. I want our family to stick together no matter what, especially with everything that's going on now. You don't know what's going to happen. When I was growing up things seemed more coherent or stable, at least to me, but I didn't have to deal with the financial part of it. My faith in the Lord has brought me through a lot, and I'm positive it will get me through whatever comes along.

After May 20th, and BSE, Wade and I were a little concerned that cows would be worthless, which they are right now, and we made a decision at the end of June to sell them all. We have a few left, some dry cows and the kids' cows and calves, but I guess we just thought we'd rather be getting in on the front end than the back end and took a chance and sold them as pairs. It worked okay. We'll buy calves and grass them in the spring or whatever, depending on the market. Cows are hard to get into when you don't know if you're going to be able to sell them. That's one thing about Wade, he's not afraid to try something different. It's been a great learning experience.

Yes, this is change. And it's big change. It would be easier to have no background in it, – say you just got into ranching as a new career and you had no experience, then it would be easier to accept changes because there would be no preconceived ideas about how things should be done. You have to think more of how you're going to market. You have to make money more than five times out of ten. In the good years, you're stretching the dollar, and when things aren't so good, in those bad years ... well, the bills keep going up. The people who make everything we need and people who are in non-related jobs – never, or very rarely, do they ever get a pay decrease. Their pay always goes up. The accountant's wages go up, the teacher's goes up, the guy who drives the grader, all those kinds of jobs always go up, but the food producing people, no. Because our country has never had to starve, food is irrelevant. Entertainment is the priority in most people's lives and that's where people spend all their money. Food is cheap in relation to other things in a lot of people's budgets.

It's frustrating when the rest of the world subsidizes producers. It's my understanding that we were supposed to cut all these subsidies out so everyone is on the same level, but our main competitors are subsidizing more than ever. We don't do anything – it's like our government really doesn't give a hoot. If you don't live east of Manitoba, it's like you don't exist. Especially right now. They really aren't doing a whole lot to help this BSE out. And maybe there isn't anything they can do, with the border closed. On the one hand, you'd like them to help, and on the other, you'd like them to keep their noses out of it. It gets to be a real emotional issue. We need to relax and make changes where we can and not get all caught up in the things we can't change.

Things are going to change a lot, with what's going on with environment and all of that. Eventually carbon points or whatever are going to come into play, and there'll be more environmental issues brought up and you're going to have to have an environmental plan for your place, and if it doesn't measure up well, then you'll probably be shut down. That's how I see it. I guess the people with the money will make the decisions, just like it's always been – when it comes right down to it, that's what will happen. Here, anyway. I don't know about Alberta or Montana because a lot of the ranches down there have been sold, the great big ranches, because they can't pay for things on a ranching income, so what happens is they sell them off to some celebrity or someone with a lot of money and then they work for them. It's probably going to happen here, it's happening here already. It's not ranching money that's coming in and buying up all the land.

I think what you have to do is you have to figure out a way you can make a living with the lowest cost of production, and if you can't cut enough, then you'll have to find something else to do. It's not going to be the same as it was in the last thirty years. It has to change and it is changing.

Ranching is a great life and a great way to bring up kids. I love this place we live on – it is like a big playground for our family. I do have to remember it is God's and try not to get attached to it or to make it my god.

LINDA FROSHAUG OUT FOR WINTER FEEDING

LINDA FROSHAUG – Bonnie Llattas Cousin

Maybe runaways are a state of mind. You have to want control in order to have a runaway. Otherwise you're just having a fast ride.

Born in 1949, the daughter of Lorene and Boyd Anderson, Linda grew up on a ranch in southern Saskatchewan. She and her husband, Ralph Froshaug, ranched south of Glentworth, Saskatchewan. Linda had four children: Corbin, Ryan, Bonnie, and Justin. She taught English in the local high school, where she shared her own unique philosophy of life. Otherwise, she was found wherever there were horses for riding or driving. Linda died of cancer in September 2003.

I DON'T KNOW if we used the word *ranch* a lot. I don't know that I felt it was my word. I read a lot and the books were always about such and such a ranch. The word *ranch* would be on the title of book covers, but the ranches described in those books seemed to be different from ours. Maybe "real ranches" should have had more horses and more bucking, more of that sort of thing. We had sheep. Sheep didn't fit in with that concept of the Wild West, and our horses certainly weren't of that type. We didn't have fancy saddles and horses and ropes. Ours wasn't a ranch, not in my mind.

I never thought about guns, so I don't associate them with a ranch. I suppose the only Wild West I might see would be a gun during hunting season when guests would come from the city and go hunting, or the gun would come out when we butchered. As a kid I always got to watch the butchering. I was very fascinated by it. I still like watching it. Some places they always made the women and kids go in the house when they were killing things. But I don't remember Dad making me go in the house. I think I was just like a shadow. Watching them die, just such a very powerful moment when animals die, actually die.

The worst thing of all is putting down horses or dogs. I guess I just want to be there and make sure it's okay, that everything is done, and it's okay, but we've had to put down a couple of horses in the last while and it's really difficult. To put horses down and dogs, too, it's really tough. Maybe it's our judgment of their intelligence. As a human, my suspicion is that we really have a high opinion of ourselves as a species and in sort of a godlike manner; we believe we are a power over other animals. We think this gives us a feeling that we can also identify with them emotionally. So we sort of put our own emotions on the animals, like "they are terrified, they are sad." I don't think that's the case at all. My knowledge of animals is that they don't have our emotions. Their grief is just different, even though it has a great intensity … this is sort of metaphoric. For example, mothering is the strongest attachment of any animal, but even a cow will only grieve for about a week, ten days, and then she is done with it, she doesn't drag it on. I'm not saying there isn't a sense of loss, but I don't know how much of it is associated with the fact that she has a really tight udder. I think we're doing the animal a disservice by putting human emotions into their lives. They probably have piles of emotions, but they're animal emotions and they're not ours, and I think it's egotistical of us to presume that we can judge them. They're far more complex than that. It's not a simple thing what they're doing out there, it's just not the same as us. That's all.

My earliest memories are of the sheep, the raising of the sheep, and being down on Rock Creek. I would be going around with Dad in the truck and I'd be standing on the seat – this was before seat belts and car seats – and Dad would usually be numbering the ewes and the lambs. A ewe would have twins or triplets, and then Dad would put the numbers on. I think that's where I learned my numbers. He'd have them all in the back of the truck – different coloured paints that is and he'd slap the numbers on the triplets or the twins. He had over a thousand sheep so we'd mark them. And then we had the bum lambs in the barn, and I really remember one little blind lamb. She died in the barn.

Under five I just remember trailing along, tagging along. Dad would herd the sheep on the prairie and then we'd go north and then we would have bum lambs. Always milked a cow. I think I milked a cow when I was quite young, probably around the time I started school I began milking. Always had chickens, we had every animal. I don't know when Dad quit the sheep, I was probably eight or nine. When I started to school it seemed like the sheep years were ending. Then we got more cattle and more cattle until I left home at eighteen, and between those years Dad built up the herd every year.

When I was fifteen, Dad was in politics and he was gone a lot. I was calving out the heifers, thirty some heifers. They were out in the hills and I had to go out there after I came home from school. It seemed like I was always too late and sometimes ended up pulling dead calves out of cows. I really remember doing a lot of the calving work by myself. I used my horse. Blaze was young then, she was only about two, and I tied the rope to the horn and I led her, and sometimes Blaze pulled until she couldn't pull anymore and the cow pulled her over backwards. Some of the horses were awfully good – like they gave that snap, the steady pull wasn't what you wanted. I know in the corrals we used a pulley system. Dad had a wire stretcher and we'd pull them in the corral with that or pull by hand if we could.

After calving we'd move the cattle to the lease, we'd round them up and move them that twenty miles. And that was down a lot of road and through a lot of pastures. And we'd do maybe a couple of trips in the spring and maybe a couple of trips in the fall. It was really interesting then because we had no horse trailers. After you rode down twenty miles, then you'd have to ride home again, and then you were falling asleep, almost falling off your horse different times, too.

I think I was seven when I made that first trip down. That was the time that Grandpa Anderson gave me the Riley McCormick because I was riding bareback. Actually I wasn't quite bareback, I had an old army saddle. I remember my Grandpa came out to the road and he was in the El Camino and he was talking to my Dad. He must have been watching me as he drove home, and he came back with this little saddle. From then on I had a saddle, and that was really wonderful. I didn't have cowboy boots or blue jeans or anything but I had a saddle.

I did lots of work on my own. I don't know how useful I was – probably pretty useful. Thinking back, I helped Dad a lot, put cattle up the chutes and that kind of stuff. Milk cows, I remember because I broke that milk cow to ride. I rode everything and I trained them too. I rode sheep, too – every animal that was in the yard, I rode. They just can't get rid of a kid. It's like catching a horse. Sometimes those horses did not want me to catch them. I'd follow them for three or four hours, and they just gave up because they knew I wasn't going away. If you were trying to catch a horse in a big pasture that doesn't want to be caught, then you just stayed behind it.

And then the branding used to take place in the summer down at the lease. We used to brand in July in those years, branding much later than we are now. Calves were born later and they'd just be smaller, too. I was pretty well an adult before I actually roped. I never did get to wrestle much, even though there were a lot of boys around. I wanted to but ... I was vaccinating years later. I don't know that any of the kids had a big role, maybe the boys. I'm sure they made an effort to get the boys to wrestle, but they didn't make any effort to get me wrestling. I enjoyed it a great deal but that was after, when I was on my own, with my own cattle. Girls could go help cook or something. I resented that a lot.

When I got into a position where I had my own cattle, then I did every job. Like now I do every job. Branding, using the branding irons, was strictly a male job. So and so would have to put the iron on. Like you had to have some great gift. It doesn't require any huge amounts of intelligence. Most brandings that I go to today, you can see a woman doing any job. Anywhere. No one pays a whole of attention. I don't think there's a big trick to castration either. There's the myth, but I think you just do it, it's a surgical procedure, one that gets done, and it's not pleasant and you know you just get it done but there's so many myths attached to it, it's unreal. I think men made them all up – they have a complex or something. Nothing too drastic. I don't want to be too dramatic about that.

Probably I didn't do much in the winter. Just on the weekends. I remember going out with Dad to feed on the weekends. Sometimes I'd get the saddle horse in and ride along beside the hayrack. Dad used a team and we'd feed down on the Hudson Bay quarter.

And I missed a lot of school, an awful lot of school. I think there were years when I managed two days a week on the average. I think my mom was very good to me. I always appreciate that – that was the best thing she ever did for me, she let me stay home from school. In the winter we were snowed in, so we were home all the time anyway.

As a kid I rode everything, and I trained them too. I started early on Honey. She was a milk cow's calf and then we were milking her. I took people double. I'd hold her tail over one shoulder and I had a binder twine around her neck and I was in between there, and then I had a little stick and I'd tap her on the ear to turn her. I could make her go anywhere, but sometimes she galloped, the way milk cows do at a little gully, they always want to go lickety-split through there, bucking and kicking, and away we'd go flying.

We didn't have real good kids' ponies, like Dad didn't have any kids' ponies. We had a big Percheron mare, Judy, but the first horse I rode was probably Napoleon. He was always crazy. He shied at everything – he shied at dead animals and he thought they were everywhere. He really wasn't a popular horse, like no hired man could stand him. Dad wouldn't ride him for any money, so I did, but that was all I had. I no sooner started riding him than I started driving him as well. I did both. I think I was just as fascinated by teams at the beginning as I was by riding the horses – the two came together. I drove old Napoleon on a stoneboat, just a little tiny stoneboat, and I nailed a little paint pail on it and I stood on it. I didn't have shafts and I'd slide up between his legs, and I'd slide off. I drove him for miles over the hills, and I'd fall off and hit rocks. Riding by myself, independent riding, probably began with Napoleon and Judy, the only two horses we had at home.

Then there was Blaze. I never did own Blaze, I just broke her. I trained her completely, but you know how you never get your name on stuff. She was part Arab and part Percheron, so she was a very ugly colt. Dad was focusing on others. I just sneaked out and I broke her and was riding her and they didn't know it. Blaze was a bit of a snot, too, but she was so fast.

Betty was mine. She was a thoroughbred and she was a rattlehead, completely. I think I could do stuff with her now – I know more things now. She ran away on everybody, but Dad didn't realize how bad she was because I wasn't going to tell him. Then my only strategy was I jumped, just jumped off. I really couldn't control her at all. Betty was hot-blooded and I guess the best thing about her, she was a jumper. The most fun I had with her was jumping. There was a big pile of lumber by the barn, she jumped that. Trees, everything ... I jumped bareback. I'd get her in a stackyard and make a bale stack, and then I'd just fly over it.

The first thing I did when I got my own place, I went across the line and got this old team and I drove them. When they were getting pretty old, I went to the PMU and I bought a couple of colts at a sale and I started them, broke them, and went on from there. I think I broke nine teams, nine teams. One by one I sold them or they got old or they had problems – one would get lame and I'd have to sell them. Some ended up for meat

We always went on the wagon trains, like when the Boundary Commission Wagon Train came through from Manitoba to Alberta. We went on maybe three sections of it. And there was a wagon train every year from Willow Bunch to Wood Mountain. And a wagon train with the Mankota people, over to Wood Mountain. School trips, for years I took kids out for an overnighter. Sometimes I still take them down to the lease, – it's twenty miles from home to the lease.

I don't know when I started thinking of myself as a cowgirl maybe when I could get into rodeo. Barrel racing. And I liked to ride fast. I was fearless, yeah, I didn't have much fear.

Somebody told me a story how I rode home over the top of the hill wide open, bareback down the bottom and up the other side. I could have been on a runaway for all I remember. Maybe runaways are a state of mind. You have to want control in order to have a runaway. Otherwise you're just having a fast ride.

Sixteen days before her death on September 20, 2003, *Linda wrote:*

Last night I read the interview. It was like looking into a mirror of my soul, especially the fine paragraph about the runaway. I truly understand that I am not on a runaway here in Regina, fighting stomach cancer. In order for this to be a runaway I would have to want to control my future. I am really just on a very fast ride.

TAMMY BURGESS

When we were married Michael gave me a saddle ...

Tammy Burgess loves life. She is exuberant about everything – her marriage, her family, her work, and her art. Born on a farm in Montana in the 1960s, Tammy is a transplanted American. She presently lives on a ranch in the Big Muddy valley Valley with her husband, Michael. They have two daughters, Tiffany and Britt, and a son, Lane. Tammy and Michael are partners in the ranch with his parents, Edward and Ferne Burgess.

IT'S TWENTY-ONE YEARS since I first moved to the Burgess Ranch. I grew up in the Big Muddy Valley, four miles southeast of here in the United States and now I live in Canada.

At school I studied all the things that I particularly cared for and I always thought I wanted to become a bookkeeper, so I studied a lot of math and bookkeeping. When I finished school, I worked in Plentywood and lived with my sister. That's when I met Michael. I loved to dance and was out at one of the nightclubs with my brothers and friends, and Michael and I were introduced. We were very surprised to find out that

TAMMY BURGESS BAKING PIES FOR A ROUND-UP

we grew up so close to each other, just four miles apart. That was one of the biggest surprises because when I was growing up, the Canadian border was the Canadian border, it wasn't something that you took lightly. We knew whose land was right next to ours, and that was Tommy and Blanche Marshall. Jim Marshall had land on the west end of us, but that was the extent of the people that we knew in Canada. As for the Burgesses, I had never heard of them. I was quite surprised to find out they lived so close. Our land is about a mile and a half apart. Our houses are about four miles apart. Any place else in the world, if you lived within four miles of somebody you would know them. You would probably have them for tea or coffee, but we didn't because this was along the Canadian border.

Michael had been away at school, but in the summertime and on long weekends he was at the ranch. When his sister was ready for school, they moved to Moose Jaw, so he finished his schooling there before he came back to the ranch. And then between the time he finished high school and the time that we met, he also went to Olds College in Olds, Alberta.

When I first met Michael I thought he was very tall. I'm only five foot one, so I notice tall people. And I liked him from the first time I met him. A lot of my friends thought, oh no, another Canadian, but I guess he was the Canadian that was right for me. I always teased him that I liked his cowboy hat, maybe that's what got him I don't know. The fact that he was interested in horses got me because I love horses. I always have.

I remember the first time when I came up to ride horses with Michael. My brother dropped me off and he went on to visit some Canadian friends, and when he came back, we weren't back yet. Mrs. Burgess was quite concerned that maybe I got hurt on the horse because she didn't know I could ride. When we were married, Michael gave me a saddle – it was custom-made for me, but by then he knew that I would use it. He's like that. As far as what it was like when I first came here, I guess I tried to fit as best I could. The first year Michael and I were married, I was out cutting hay, raking it while Michael baled it. I've always been able to help with the cattle outside, as long as I wasn't pregnant. I'm sure Michael expected me to pull my own weight and I wouldn't have expected anything different. I asked him one time what would have happened if I hadn't been the type of person that liked to ride horses and work outside, and he said we probably wouldn't have been married because he knew what kind of woman he was looking for and I just happened to fit the bill. He's very happy when I go outside to help him and I still do a lot of the calf checking in the spring. To me, this is a family ranch and I'm part of that family.

Right from the beginning Michael and I knew we wanted to be able to take over the ranch someday so that his parents could retire. We wanted to have children. And be able to keep the ranch viable so that they could have the ranch when we retire. We talked a lot about the different things that we wanted and how we wanted to accomplish them. I can't remember when this first started, but when we were dating we talked about what we could see ourselves doing in the future, and probably because of those talks, Michael knew that I liked the ranch, that I wanted to be here.

When we first got married Edward was running the ranch – it was his ranch to run and we were the hired help. As Edward grew older he allowed Michael to have a little more say because he knew Michael would eventually be taking over the ranch. It's to the point now where Michael is running the ranch basically, and Edward helps out where he can. We don't feel that's he's the hired man by any sense of the word, because he's the one we rely on to get another opinion when we need to kind of bounce things around. Our son is being raised the same way. He's working for us right now, doing the work, but it's not like he doesn't have an opinion and we try to listen to him and make judgments on whether they're valid or not.

When the kids were small, I was fortunate in having a mother-in-law and father-in-law in the same yard, so if there was something that had to be done outside like riding or calving or whatever, they were able to take care of them. I didn't have to worry about what was going to happen because all I had to do was call on Fern.

We always took our kids riding. I remember taking a ride after our son was born and he was tied to my stomach, he was in a snugly. As soon as I got up on the horse, then he was in a little bag tied to the front of me and we went for a ride. The girls always rode. Tiffany used to go out on the hillside in the summer. We had one little pen of sick calves and we'd let them out in the daylight to graze, and then she'd go along with her quarter horse and bring them along to put them back in the pen. I imagine Tiffany would have been five or six. She was riding a really, really quiet horse and he was a good babysitter. Britt started on a pony because Michael's sister had a pony that she brought out, and she rode along with us, and Lane was the same way with the same pony. I remember him riding when his feet didn't even reach the stirrups.

In '86 I was pregnant and it was a very icy type of winter, and Michael was concerned if I went outside around the yard like I had been doing, that I might slip and fall. I decided I would take up a hobby so I chose to do crafts. And when it got to the point where doing crafts just wasn't challenging enough, I changed to the medium of acrylic painting on canvas. I've never taken a course, it's just all in my natural abilities, I guess. The more challenging I could make it, the more exciting it was. Michael has always been a great supporter and if he found a piece of art work that he thought I might be interested in, he would buy it for me for Mother's Day or my birthday or whatever was coming up, just to see if I would like to try something like it. I remember one gift was a painting on a feather, and I do a lot of that now because I thought it was very intriguing, very challenging to be able to do that.

Most of the artwork I do is of a Western theme. I do landscapes – the Big Muddy is unique. People come to Saskatchewan, or you tell them you're from Saskatchewan, and they say, how can you stand it, it's so flat. Well, a lot of people don't realize just what Saskatchewan has to offer in the line of rugged, untamed areas. We've had lots of people here who are just kind of taken over by the beauty and the ruggedness of this valley. Unless they've been here, people don't really realize what it's like.

Mostly I paint horses and cows in the landscape on our ranch. I was asked one time how many paintings I have done, and I couldn't even hazard a guess because I don't really keep track of them. When they're sold they're not here, so they're taken off the list of what I have. Sometimes I wish I hadn't sold "that one," but knowing that someone else is enjoying it is payment too, not only monetary payment, but the idea that someone else thought enough of it that they wanted it in their home makes me feel real good.

I do a lot of wildlife painting, too and don't always do them on canvas. I do them on rock. We have sandstone here in the valley, most of the hills have some kind of a sandstone deposit in them, and I've chosen to use that as one of the things I've painted on.

My gallery started as a need for someplace to hang my paintings other than in the hallway of our trailer house. In the middle of a meal, we'd have people going down the hallway looking at the paintings. That was in '94. And I decided I would build myself a gallery. We were doing some renovations to the original house on the ranch. We needed to take the sun porch off it, and Michael said, if you want the porch, you could take it and use it for your gallery. So I moved it into the back yard and put a back wall on it, and I did all the electrical work myself and I had a beautiful little gallery in the back yard. By '98 it was decidedly too small because I was having so many people coming in to see my art. One day I had two big forty-two-passenger buses come in, and they were trying to see in that little gallery, which just wasn't big enough. We had another house in town and it was falling apart. If someone didn't salvage it soon, there wasn't going to be anything left to salvage, so again he told me if I wanted it, I could have the little building. So I took our son and went up and tore the building down, stick by stick, and brought it home and rebuilt the new part of the gallery, and we had the grand opening in June of '99. And it was a lot larger, a lot brighter, a lot cheerier, and we can at least get more people in it now.

I usually paint in the house. I do framing now, so the original gallery is used now for storage and for framing. But I still like my kitchen table when it comes to painting.

Michael and I lived in a trailer house for twenty years. Then we decided it was time to build a new home, so I looked through every book I could for floor designs for a house that we wanted. We wanted something that was open and small. We didn't want a really big house. We wanted to have a big feeling, but not really all that big. We wanted it to welcome people that come to visit. And I think we accomplished it. It's everything that I wanted in a house and a lot of the things that Michael wanted.

A loft bedroom, I've always wanted a loft bedroom. I've always wanted a big covered porch in front of my house. Michael and I talked about building a log house, but when we found out what the upkeep of a log house is, we just decided that we wanted a house – that our dream home was something that would not keep us busy the rest of our lifetime. So we chose to build our house out of concrete – a lot of people think that sounds strange, but it's actually concrete, the walls are twelve inches thick. On the inside we chose log siding, so it has the effect of looking like logs, but it has the durability and soundness of the concrete. I got my big front porch with my custom-made railing with our brand in it. I just wanted a non-typical ranch house. Hopefully I'll never see another one like it. I wanted ours to be unique.

Parts of the house were a challenge. I went to Regina to learn how to build a concrete house. I designed the house myself, with those twenty-foot high ceiling beams. I left it up to Michael and the carpenters as to how to get them up there. We didn't have a crane, we couldn't rent the equipment needed, so we had to make do. We used the front-end loader on our ranch tractor and put the beams into place. After they were put into place, the ceiling was put on. The carpenters that built the house were kind of baffled how the house was going to be built because they had never built anything like this before, but I think they were quite surprised at how well it turned out.

We live on a ranch. I don't see where there's much sense in having a house like you would have in Toronto out here. If somebody walks in my house with their shoes on, I don't want them to feel bad because they got mud on the floor. We're ranchers and we get mud on the floor all the time. We can handle it, we can clean it up, and it's not a big problem. I didn't go with a lot of carpet in the house for that reason. When we have roundup in the fall, all the riders, cowboys end up having dinner in my kitchen, and I pull out the table as long as I need it and I feed them all, and I don't really care if the floor gets dirty.

There's a Western influence all over the house. I had boards that were left from the house and I don't like throwing anything away that could be used, so I designed the clock and the table and the floor lamp and the plant stand so they would all match the house. I love horses and you can tell that by the horseshoes that are hanging on the walls. There's five horseshoes on the wall and they represent one for each one of us. And they are [turned] up, so hopefully we'll always have good luck.

We love the land. We had a seismic outfit go through here a couple of years ago, and it was painful to see what they were doing because there were places that had been virgin to a tire track. You know, they say, that land is protected, most of our land is in wildlife habitat. We love this land and we don't want anything to ever happen to it. My father-in-law, in fact my husband's grandfather got this ranch in '37 and we've had it ever since. And we take care of it to the best of our ability so that it will be here for another generation. We need it to be productive because we have cattle and if they don't do well, neither do we. If our cattle can't get grass, if they can't get water, we don't do well. I mean, we protect it because it's our livelihood. We've also found people don't realize when they're out on a Sunday drive, they can't just drift through and leave gates open and think that they're not going to have an impact on what's going to happen later. For years we've been going around shutting gates because people seem to think they can drive through.

In the spring if we don't get enough grass or if we don't get enough rain, we realize that early and we don't try to keep as many cattle as we normally keep. We want to see grass standing there at the end of the season – it catches the snow and can do the most good. If you overgraze it to the point where there isn't any grass left, then you're just going to have the snow blowing away, there won't be anything there to catch water the next year. Conservation and the grass is very important in this area because what one person thinks is a tiny footprint in twenty years is going to be a big hole because of wind erosion. This land that we ranch on is very fragile – you'd think it would be rough and tough type of land, but it's not. It's a very fragile area. Whatever we do is going to show up ten times over in the future. Better watch what we're doing all the time.

I don't work off the ranch. I have the stability of not having to. I know there are a lot of women that have to get jobs off the ranch. What really bothers me is when they say something like, "Well, you don't do anything?" That's not exactly true. Just because I don't work off the ranch doesn't mean I'm not working. I'm a stay-home mom. Yes, I am a stay-home mom; I'm raising the next generation. I may not work off the place being an accountant, but I do all my husband's bookkeeping and I know exactly where we are financially. When my husband needs someone to come out at two o'clock in the morning to help him pull a calf, I'm there by his side. So when somebody asks me if I work, meaning do I work off the ranch, no, I don't. I have enough work to keep me busy right here. And in that way we don't have to have a hired man. But it really bothers me a lot when somebody asks me if I work off the ranch. It's as if I don't work on the ranch when they're asking something like that. That really irritates me. We have a twenty-year-old daughter who's going through college right now and expects to be out in the workforce. I don't expect her to be a stay-at-home mom if she doesn't want to be. It has to be something that you want to do. I never was driven to get a job off the place. My job happens to be here.

Even so, I don't think I would be able to get along just doing the same old thing every day. I enjoy my painting very much and it's very pleasing to me to be able to sit down and think about something rather than the everyday humdrum of life. This is something that I'm doing for myself other than ranch work. We've accomplished a lot in our ranch life, but my artwork is special to me. It's the one thing I can share when and if I want to. I wind up sharing it a lot because Michael appreciates the artwork.

RUTH PRITCHARD

It made me tougher.
I had to pull my own weight.

Ruth was born on a ranch south of Mankota in the 1950s. She still lives on the same ranch with her husband, Kelvin, and they are partners with her brother, Otto Rausch. The ranch is a large operation with a home place, as well as a camp at a lease several miles away. Ruth and Kelvin have two adult children, Riley and Sheena. Riley works on the ranch, while Sheena is married, lives in Mankota, and works at the vet clinic. Ruth's creativity is revealed in the landscaping she has developed around the main buildings. This interview was infused with laughter, especially when Ruth told of tough times.

WE HAVE ABOUT a township of land, most of it is lease, in about three different groups of land and we mostly raise cattle. We raised buffalo for a little while, but we went out of those – they were no fun. Mostly we sell yearlings. With our operation we have to calve later – we don't have the facilities for earlier calving to get the weight on, so that's why we went to yearlings. We can make up time.

RUTH PRITCHARD ON THE ATV

My dad homesteaded seven miles from here. My mom came originally from Russia and then married Dad. Later we moved to this ranch. They were older when I was born. There was Erna, then Otto, and I am the last. Dad ranched with his brothers until he retired. Then he moved to Mankota and that's when he gave the ranch to Otto. Otto tried to ranch by himself for a while and found it was overwhelming. That's how I ended up coming back here. Otto and Kelvin and I own it now. Our son Riley is working his way in, too. We had a pretty good deal, basically you could say it was given to us. We could never have made it without Dad.

Decisions. The three of us always talk, I am never left out. Sometimes I wish they would make their own decisions, like buying tractors. I'm not into tractors, but they want me to be there – my opinion is regarded. We set goals and we figure out if we can afford something and when, and in that way I am always involved. The decision part is the hard part, trying to make everything work out. I try to stay away from the bookkeeping. Kelvin does all of that. It gets so complicated with all the partnership stuff, too.

When we got the ranch, we added on Eddy Lockhart's land and a little more here and there. This is not farming country. Nothing grows here. Just grass. Native grass. We're fortunate that we have water – we have some creeks running through the lease and on Eddy's there are some springs. We leave the cattle out as long as we can in the fall. It depends on the weather. When we wean, we bring up our calves and leave the cows down in the pasture. Then we calve out down there about April first. Otto and Riley go down to the camp for about a month and calve out the cows, and Kelvin and I keep the heifers up here. We have to buy a lot of feed.

One spring a snowstorm was so bad down south we had to go in a four-wheel-drive tractor to help out. Kelvin and Otto were out on horses finding these calves. I remember I was out with Eddy Lockhart in the tractor and the cab was full of calves, and one was down with its legs under the clutch and Eddy couldn't shift and I was down trying to get it out. The tractor kept cutting circles next to this steep bank and we kept sliding down. We had two days of terrible weather. When it ended, I was digging calves out of snow banks and my eyes were so sore I stuck Kleenex over one and tried to get by with the other. Finally I couldn't stand it and they sent me to the house. I guess I was snowblind. Calving is always stressful, but it's amazing what you can put up with if you have to.

It's a totally different way of living now. I remember when I was young we had Sundays off. Now, you'd think it would be easier with all the machines, but I can't remember the last time we had a Sunday off. We have to earn more now to pay the taxes, the lease and stuff, so it's a full-time job. It really is. And my family worked hard – they worked their six days, but at least they had Sunday off. Theirs was all handwork. They didn't have the machines like we do, stackers and all. It was all square bales and a team of horses. I don't think they fed like we do either – their yearling weights were way lower than ours are now. Now if you came up with the weights they did, people would just kind of laugh at you. I think they had hardier stock back then too, because they didn't penicillin everything like we do now. Nowadays you can't afford to lose anything – back then, well, they didn't put the sick ones in the barn like we do now.

My mother never had much of a role in the decision-making on the ranch. She had a huge garden, and she milked cows and had chickens, all that, but as far as riding, no. She would be the one cooking at branding. Bonnie Loewen and I were the first girls that went to all the brandings. We didn't go to the kitchen and work. We went to the corral and gathered in the morning, and then we might go help in the kitchen. We were the first girls around here that did both. Later we got to brand and everything.

I was about thirteen when the work really started. They didn't want kids around the corral because we might get hurt. At thirteen I got my first horse and I had to break it myself, and from then on things were a little different. Basically I got on and I rode. I didn't spend hours and hours in the corral taming it down. We'd snub the one we were breaking up to another horse and then they'd turn us loose. My brother would help me. I had a buckskin and we were going to a branding. I said to Otto, "You'd better try her this morning." And he said, "Aw, she's got to buck you off three times, and then I'll try her, but three times first." That's how it was around here. It made me tougher. I had to pull my own weight.

We continue to break horses. I've got one on the go right now. I'm older, the ground seems a lot further away. Now I run them around the corral a few times. Back then I just got on and if I fell off, I got up and got back on again. I like that feeling of being able to get on and say, this is my horse, this is the way I like it. I don't want someone else's broke horse. This way I'm not so quick to criticize.

We used to use a team all the time in the winter, but now we're into round bales. We still have a team and we raise a few colts. With square bales it was handier. You know, you just go at a slow speed and drop the bales off. Now it's always a rush job around here. We eat and then we're out again, and the men help with setting the table and the dishes so we can get out faster.

Before I had the kids I was out constantly, and then things changed.

I was married when I was seventeen and I wanted to make sure the marriage was going to work, – I wanted to make sure I was ready to have a family. I remember being huge and carrying slabs, putting up windbreaks. I think it made everyone a little nervous that I didn't slow down. Then when the kids came, they were bundled up and they went out with us too. I mean they never had babysitters – they went to grandma's, but mostly we took them along.

I think Kelvin and Otto appreciate me. I know they expect a lot, maybe a lot more than others, but they appreciate me. I'm not a housekeeper. I'm outside, vaccinating, sorting, and so on. The day before a big job like sorting and preg checking, I'll try to do the cooking. Then my mother-in-law will come and set it out so I can be outside working.

At branding time we get to see people we haven't seen all year. We start gathering about seven and then we separate the cows and calves – that is, the women do that while the men castrate and brand. And the women vaccinate the cows for BVD, usually about six hundred head. Otherwise we'd have to have another day. So when we get done, we can all sit down, have a beer or two, eat, visit, and a few pranks are played.

Leisure. Would it help if I told you we bought a boat but we never get to use it? We haven't taken it out for three years now. When the kids were young we decided we should have this boat for water-skiing. And for a while we had a couple of days of water-skiing at Saskatchewan Landing. If we had five days holidays a year, we were in the money. And then we went on about three wagon trains. I can't believe we took that time off. Christmas and brandings, the neighbour's brandings – that's about the only time we take off now.

We never expect to be rich. We've got to like what we're doing because any money we make goes back into our ranch. You live year to year, or we do. We're not going to make trips to Hawaii or whatever. Every day, we have to love what we do. The way things are going right now, so many ranches are in trouble. Unless you're well established, it's kind of scary. I know there are lots of people who don't like their lives, but I can honestly say I love to live on our ranch. My whole life has been right here, twenty miles from town.

SHERRI GRANT

A ranch is a business.

Sherri Grant grew up on a farm near Edam in the 1950s. After completing a degree in home economics, she worked in rural Alberta before marrying Lynn Grant in 1975 and moving to a ranch at Val Marie, Saskatchewan. The Grants are part of a family ranch, and for Sherri the emphasis has always been on communications within that ranch. Sherry is a vibrant ranch woman, set to approach the twenty-first century with twenty-first-century technologies. Sherri and Lynn have four adult children.

THE PRIMARY THING from my education that I think has been useful – well, two things – one, the communications aspect and second, the job opportunity that it gave me to work with rural people and learn more about rural people in other settings.

It's interesting now because people ask me where I live or what I do or whatever, and I tend to say we farm and ranch. I tend to never use the word *ranch* by itself and I think it's because I'm uncertain as to what other people think a ranch

SHERRI GRANT PHOTOGRAPHING WILDLIFE

is. Maybe because when I was young, a ranch was this great big glamorous wonderful place and it's not like that. A ranch is a business, just like all other agricultural business. I don't want to give people the wrong impression.

We have a cow-calf operation. My husband and his brother have been developing the business that they operated with their father. It is now triple the original size. Their father, Stuart Grant, had three sons and his goal was that each of his three sons would be able to operate the farm-ranch if they so chose. So that is how they set it up. One brother operates on his own, and my husband, Lynn, and Dean operate together. Two families run our operation, it always has been. Because of that it's probably bigger, or seems bigger, because we think about the operation in its wholeness, not it's half, but the reality is that it supports two families. We run about seven hundred cows. We background calves – we have a feedlot in the winter and I can look out the kitchen window and see the lot where we feed the calves. We usually buy calves as well, so we winter anywhere from 1,000 to 1,500 head of calves. Then we put those yearlings out to grass the following summer. Our cattle graze on about 20,000 acres of native pasture and 6,000 acres of tame pasture. Our usual program is to sell yearlings when they are about 900 pounds. That is somewhere between early July and mid-September. The open heifers are sold in September or October, then some years we sell bred heifers the end of November. However, this year with the BSE, the world's all changed. This year we're retaining ownership, so that's another story.

We also grow and harvest all our own hay. We have 500 acres of border dyke irrigation and some flood irrigation. We also bale dry land hay and green feed. Right now we're in the process of doing a second cut on some irrigated alfalfa acres. We also seed annual crops, and in the past we primarily had it custom done. Until last year we were seeding about 2,000 acres. We continuous crop to prevent erosion from the sudden hard rains and the wind. Last winter we purchased some additional farmland, and this year we seeded about 3,500 acres. This year we got involved in renting seeding equipment and doing our own seeding, buying a sprayer to do our own spraying, harrows and a Val Mar to do that work, and it went on and on. We invested in a Macleod Harvester, which is a new harvester, and did our own harvesting this year. So this year has been an extremely steep learning curve in terms of our operation because we've had a lot of new things to learn about that normally we weren't directly doing. That's sort of us, the big picture.

I tend to work the communications centre at the house, the phone and the two-way radio – sometimes both at once. I also do volunteer ambulance work, which means I have to be within ten minutes of the ambulance, so I have to be near Val Marie. Along with my community involvement, my work on the place is the accounting, dealing with banks and the accountants, tracking all finances, and what goes on in what field so we can do enterprise analysis, those kinds of things.

We have raised four children. Our goal as a family is to create an operation where our children have the opportunity to become involved. We've seen other families who didn't expand their operation and their kids left because there wasn't enough money for them to be supported by the farm or ranch. Then when the dad wanted to retire, the children had lost interest in coming back. Our goal has always been to raise our kids to be independent, to be able to make their own choices, and to have a diverse operation so that if they chose to be in agriculture, it would be a good business for them and a good place to raise children.

Our oldest is Amy, she's twenty-four. She has a degree in commerce. She's now working as an administrator at Red Coat Feedlot. She's married to Wayne Andre, who farms nearby. Amy and Wayne live twelve miles north of town. Our next oldest child, Logan, went to Olds College and trained as a heavy-duty mechanic. He's just finishing his journeyman's certification with Brandt Tractor in Medicine Hat. Logan hopes to come back to the family operation. He was back this spring for a week helping to seed and back again this fall helping to harvest. Brenna is our third child. She's twenty and she's in her third year taking ag economics at the university in Saskatoon. She came home the first of May and managed the cattle herd for the summer. She took care of calving out the cows, took care of the yearlings, grazing program. She put out mineral, moved cattle to new pastures, fixed fence, and checked water. Our youngest child is Morgan. He graduated from grade twelve in June and he likes to do the haying. Morgan has just gone off to university, so he's taking engineering in Saskatoon. So now Lynn and I are on our own again after twenty-five years. It's a little strange – cooking for two is a challenge.

In the future I hope to see more of our family becoming involved in the operation. It will mean more people involved and that's a challenge. Growth is change and change is not always easy. I'm looking forward to a larger operation, possibly more diverse and therefore, more financially stable. I see my role as not a lot different, but maybe more shared responsibility.

When I first moved here, one of the biggest things for me was loneliness. My husband was outside farming all day, so either I could go help him or I had no one to talk to. I went from having my own income and my own money, to none. So I got a part-time job working as an aide in the school. I got to know one of the teachers really well and she's been a good friend ever since. But it took me a while to get to know people in the community. I had to get involved in the community.

I never wanted to live near town, but Lynn's parents lived on the edge of town and we ended up building our house in the same yard. I was very concerned about raising the children so close to town. I would insist that my children were home directly after school. But I've come to appreciate that closeness – it meant that they could participate in after-school activities and I didn't have to do the driving.

Geographically we are somewhat isolated, but the reality now with the Internet and telephones is I can talk to anyone, anywhere, anytime, very economically. For our business the inconvenience is the distance to travel on poor roads for repairs and supplies. We often have things mailed since there is no bus service.

In our operation today Dean, Lynn, and I are the decision makers. A year ago we went through a process with our accounting firm as a ranch unit – Dean and his wife, Danielle, Lynn and I, Amy, Logan, Brenna, and Morgan – and discussed what our goals were and how we wanted to proceed. What's different for us is that it's two families working together. In terms of the operations, there's more people – more people to work with and more people to communicate with. That's the biggest difference – making sure everyone knows what's going on, that everyone has the opportunity for the input. Decisions take longer to make.

One of the benefits of working with more people is that we do have a chance to take holidays. Lynn and I went to New Zealand on a grass tour in 1994, and for our anniversary the six of us flew to the Dominican Republic. On my birthday, which was a week or two ago, we girls went to Moose Jaw to the Spa for the weekend. I took my two daughters, my mom, my sister-in-law, and a few friends. We stayed at a bed and breakfast one night and did the tunnels and various things, and we went to the Spa the second night. As you get older, you know, it seems like the "maintenance program" takes a little longer, a little more effort, and so some of us girls try to exercise regularly. We try to get together every day and walk for an hour and forty-five minutes. But if we average three days a week, we figure we're doing good.

You know, I'm really lucky because as long as I'm not working on a deadline, I can choose how I spend my time. I am currently a director with Canadian Western Agribition, on a Lay Supervision Team with our church; treasurer for the local UCW, treasurer for our Emergency Response Team, as well as other involvements. I probably spend a hundred hours a month doing management. I can choose which hundred hours I use. So I can say, "Today they're moving cattle over here and I'm taking lunch to the field." Or I can say, "This morning I'm going to take photos." This last branding I didn't have to be in the corral – I got to take the pictures.

Sixteen years ago I got a new camera. I kept hearing people say how isolated it was here and how there's nothing to see and nothing to do. Of course, I got a little irritated about all that, so I started taking pictures of wildflowers, mostly because they hold still and you can find them. I never had the patience for deer. So I took pictures all one summer and it happened to be a summer when we had a lot of moisture and we had wildflowers everywhere. I had pictures of over a hundred different native species of wildflowers in our local area, basically on our own land. And I put them together and actually Grasslands National Park has used them for a number of years. It was proof to me of the beauty of this area, the stunning beauty.

Grasslands National Park was formed adjacent to some of our land – actually some of our land is inside the Park boundary. Through Parks Canada it was decided to retain a large area with plants and animals that are associated with native prairie because in so much of the world, native prairie is disappearing, has been broken up, divided into small chunks. The ranchers here had maintained this native area and had done an excellent job of taking care of it, keeping it vibrant and alive, so the federal government chose this area as a place for a grasslands park. After many years, it is now actually gazetted, and a real national park. A Friends of Grasslands has formed and I've been involved with it, and I participated in organizing and editing the field guide for the Park. The Park is a neighbour but not in the sense that I fix this part of the fence and you fix that, or a neighbour in that your animals get into my side and I just put them back, or if yours get into mine, you just put them back, or we go fix the fence together, those kinds of things. The Park has specific rules about access, rules about doing things that aren't equal on both sides.

The Park means fewer agricultural people because that area is not in agricultural production. It's basically set aside. But there are more people in the community because of the people that work for the Park. That is a good thing. It means the community has more knowledge and expertise from different areas than it had before – we have researchers, we have biologists, we have wardens, people with different backgrounds, and that is good for a community. They bring more energy to the community.

In terms of the future of ranching, they'll have to be larger or more diverse or they're not going to survive. It's not like farming with larger machinery. As ranches get larger they tend to become more labour-intensive. And so I think ranchers are going to come up with new ways to have larger operations, but we're also going to be looking at ways to reduce our labour requirements. How we will manage, I am not sure. While we are responding to the needs of the ranch, we do not want to lose the cultural traditions. We still brand by what people call the old-fashioned way, – we still have a group of neighbours in, corral the cattle, sort the cows from the calves, rope the calves, throw them, just as the whole process had been done historically. In terms of labour effectiveness, the hours to do that and to travel to all the neighbours' brandings, it's probably more effective to buy a calf table and do batches of a hundred calves when you're ready. I think ranches are going to be looking at what's important historically and what's meaningful socially as well as financially. For people in ranching, for the children, you need to have a community, a school. It's people that can deal with other people that are able to accomplish their goals. More ideas and more input creates better results, and I think to accomplish that, operations are going to be larger, more complex, more diverse.

At the end of the day, ranching is a business and to survive as ranchers, all the business aspects must come first – production, marketing, finances, and human resources. Then, when all the business management is taken care of, if there is any time left over, it's a great lifestyle.

SANDY HORDENCHUK WORKING
WITH A PROLAPSED COW

SANDY HORDENCHUK

Recreation? Well, baling, fencing, trailing cows ...

Sandy (Wenaus) Hordenchuk was raised on a mixed farm near Verwood, Saskatchewan in the 1960s. In 1976 she graduated as an animal health technician. Shortly after, she married Jack Hordenchuk of Wood Mountain. They worked together at building up their ranch until Jack died in an accident in 1993. To meet life's challenges and to raise her three children, Sandy has always emphasized education, hard work, and teamwork. While most people live twenty-four hours a day, Sandy manages to squeeze in about thirty-six.

I WAS ALWAYS interested in veterinary medicine. I applied to a biological sciences course at Kelsey and took sciences for the first three semesters, and then the fourth semester we went to the Western College of Veterinary Medicine where we got all our practicum in small and large animal medicine, and surgical technique. I really enjoyed the nursing aspects of it. In 1976 I graduated. I worked as an animal health technician at Dinsmore, Saskatchewan for a short time, and then I got married, lived close to Wood Mountain, and worked part-time for the Assiniboia Veterinary Clinic with Don Wilson and Dan Moneo in the spring, mainly for bull evaluations [semen testing]. I was also involved in collecting semen from the Piedmontese bulls when they first arrived in Glentworth. I also used to bleed cows for Bang's disease. Then I think it was 1989 when I took an emergency measures technician course because I wasn't working that much as a vet technician.

I worked at Wald Ambulance in Assiniboia part-time, and after that Don offered me a full-time job at the clinic because Dan had retired. I worked full-time every day Monday to Friday until three o'clock. Then I worked casually with the ambulance. I continued to do this until 1998, along with the farm and ranch work.

As a kid I had lots of different pets. I loved my horses, just liked working with animals, and did chores with my dad when I was a little kid. Had my own little pails. I used to haul chop to the calves right behind my dad. Let's see, when I was eight years old, we moved from our old farmyard up to my Grandma's yard, and I think it was about two and a half miles, but yet we still wintered our cows at our old place. I used to ride with my dad every chance I could down to the old farm. We'd get to the old house, stoke up the fire in the house, and have a cup of coffee and warm up, and then we'd do the chores and come back in and finish the coffee, put out the fire, and ride home. I did that for many, many winters. I remember getting so cold, my feet were so numb, but I wouldn't tell Dad because I was afraid if I did he wouldn't let me go the next time.

And we did lots of work with cattle on horseback. We had a summer place down at Willow Bunch with three sections of grass, and we'd trail them from Verwood down to Willow Bunch every year and back, do that trek in a day.

Then I met this guy in 4-H. That was Jack, back in 1976. And we were married in October 1977. When Jack and I first got married, we had only a few cows and a little bit of grass, so we bought milk cows from his grandfather, Bill Tonita.

We had anywhere from five to seven milk cows and we had a cream quota. I milked cows and we sold cottage cheese, and milk and lots of cream. We fed pigs and always had eight to fifteen pail bunters to feed. Jack worked for Health of Animals and was a brand inspector in Assiniboia, and I worked part time with the Veterinary Clinic and for the stockyards.

We had three kids, Jody, Darby, and Baily. Poor little Jody, I used to drag her out to the barn in the morning and milk cows. It was especially tough for her at harvest, haying, and seeding time when Jack was out in the field. I got her a goat – its name was Nanny – and I cleaned out this one stall, and I used to put Jody in there in her stroller with the goat and the goat would entertain her while I milked the cows.

Recreation? Well, baling, trailing cows, fencing … and Jack liked to rodeo so we used to go to a few rodeos – he liked to bulldog. Now that I think about it I can only remember one holiday and that was to go to Swift Current to the fair, the summer before he passed away. We went to Swift Current and spent two days at the fair, water slides. Two days. I guess we just enjoyed ranching and farming so much that each day was special.

We had very good times. I actually ache to do that work again. I miss my milk cows.

I used to like sitting down and think, and maybe that's because it was family time. Everybody went to milk, did their chores. After Jack passed away, I guess I had to spend more time working at the clinic and I just got busier with the farming.

Now I have the same love for my own business. Always want to try to make it better for our employees, and clients and the patients. In 1998 Dr. Bonnie Brandt Hughes, a lady vet I worked with at the Assiniboia Vet Clinic, and I decided that we would venture out and start our own business – the Animal Hospital of Assiniboia. We had very little money, as she was a new graduate from vet school. No money, no. I put up a quarter section of my ranch land and she put up her soul, and we borrowed fifty or sixty thousand dollars to buy equipment. We rented a building. Within a year the business that we acquired was enormous and we had another opportunity. Rockglen had a veterinary clinic and they were looking for somebody to run it – we just put in our bid and thank the Lord we got it, and we've made a growing business down there. Now we have nine employees between the two clinics. In July 2002 we purchased an old machinery Quonset on the north side of Assiniboia, and again we scrounged to get enough cash to renovate our building.

We have lots of clients, and because of them we have been very successful. We have a gigantic small animal clientele, clients as far away as Swift Current, Moose Jaw, Mankota, Rockglen, Ogema – we've reached even as far as Weyburn. We also have a large large animal clientele. Dr. Bonnie's really a very dedicated and knowledgeable large animal veterinarian. We have one, two, three vet technicians, one agricultural productionist, two office managers, of which one manager is a legal secretary, Natalie Giraudier. We have two veterinarians on staff, and actually we have this amazing man that comes as a locum once in a while. He does all our orthopaedics, our embryo stuff, and lots of horse work for us.

I think there's a few clinics owned by women, not solely owned by women though. I don't think many technicians own a veterinary clinic. I don't know how well thought of I am in the veterinary world. Ten years from now? I have a dream. If Bonnie and I could find the manpower we could have at least two more satellite clinics, one to the west and one to the east. It's really hard to find veterinarians that want to come to rural Saskatchewan. You have to put in long hours – it's the time. When calving season started and Bonnie and Natalie and I were alone running our first little clinic, the hours were enormous. Enormous. There were some days Bonnie was up all day, all night, all day – she would grab an hour of sleep here, an hour there.

Actually in the first month of BSE, we've brought in more money, in bill paying. I don't know what that was about. The spin-off – even from our reps coming around, some companies haven't put out fall promotions yet, their reps don't know why. Our Tuesdays and Wednesdays here in Assiniboia aren't as busy because of the stockyards closed. No, no layoffs. But instead not a lot of stock on the shelves. So if we use "three of" then we order "three of," not "thirty of." Things like that. I know there are people who come in and their cattle have this or that, and they don't want to treat because the cattle aren't worth anything.

There are a lot of very scared producers. I'm a producer too. I don't like my bank account being empty. I feel that we as Canadians are very laid-back people and why don't we find some new markets, like for our cows. I think they're going to open the border up for thirty months-and-younger cattle. I'm holding my hopes for it this fall or early spring for sure, so our calves can go across the line. But as far as the cow crop, I'm sure the cow crop will be a while. I think we're going to have to find a new market for our cows. Hopefully other countries will take cow meat. Why don't we sell live to Russia? Why don't, instead of giving nine million or two million dollars aid to Africa, why don't we make beef jerky, or canned beef, and send it over? We have products, why should we send our cash? We could be subsidizing ourselves with the cash and sending the cows over in a can, so I think there's a lot of markets that are untouched yet.

What about the West Nile? Oh yeah, we've sold thousands and thousands of doses of West Nile vaccine for horses, and we've had to put some West Nile horses down. Some of them survive – then if they get over it, they become immune. I think like any virus, once they get bit and get infected, they may get over it. It's like the old encephalitis days – remember back when we had the encephalitis, how many people lost horses with encephalitis, lots. I think this is just another flip of the card. We haven't seen any vaccinated horses go down with West Nile yet.

My relationship with the ranch now? Weekends – weekends and evenings. I love helping with calving, round-ups, haying, riding, checking cows, bulls. This year I had a week off and I stayed home and got a bunch of stuff done. Cleaned my basement and went riding.

Can't forget the ranch hand, Camille. Yeah, Camille is a great guy. He's a very good cattleman, works hard, and we appreciate him a lot. In the last ten years we have doubled the herd – we have about two hundred cows that we calve out, plus we feed lots of calves in the fall. Anywhere from 250 to 350 to background, and then we'll grass fifty to eighty of them. Actually I sold some yearlings last Wednesday, anywhere from 85 cents to $1.09, about a hundred dollars off.

All we can live with. I know we prepared ourselves for a long dry haul and talked to the bank. As long as people can meet their interest payments, they're not worried about the payments for the year. Farm credit is rolling some of their stuff over. Teamwork is a wonderful thing.

I feel very lucky to have had a dad that taught me to appreciate animals and nature and to respect people, and to have had a husband who worked with me as a ranch partner. We raised three children and we taught them to appreciate animals and nature and people, too. It was unfortunate that we lost Jack so soon, and there isn't a day we don't think about him. The summer after we lost Jack, Camille came to work for us. He stayed on and he helps us to continue as a family. I have to give a lot of credit to my dad and mom, the kids, Camille, and to my great old horse, Charlie Brown, who made many safe trips to the calving pasture with me.

JILL MASTAD

Being outside, that's where I like to be,
that's where I'm most at peace.

Jill Mastad grew up on a family ranch south of McCord, Saskatchewan. Born in 1981, she was the fifth child in a family of seven siblings. In the spring of 2001, Jill completed a veterinary technology course in Saskatoon. She always imagined a future on a large ranch.

WHEN I WAS growing up we did all our own horse training. Actually I've been training colts for other people since I was thirteen. I rode some when I was twelve, but then the summer when I was thirteen, that's when various people had four or five horses to train. I've trained probably fifteen. My oldest sister, Carla, decided she wanted to be a horse trainer first, so Dad took her to a clinic and she learned and she passed it on. I try to do it as gently and as calmly as possible. I'm sure I revert back to the restraint way sometimes, but I'm always learning. I hope I can get where I don't have to restrain at all.

JILL MASTAD IN THE ARENA

First I take the horse in a round corral, just me and the horse. I don't put a halter on them or nothing. I chase them around until they're ready to stop and face me, and I then go and stroke them as much as I can and then just gradually introduce them to other things and back them out. I do a lot of stuff like leading them by their feet so they don't panic if they're ever caught in wire. I gradually put the saddle on. W – we don't usually ride them with a bridle the first time – and then one of my younger siblings gets on. Then I just chase them around and eventually turn them. Then we put the bridle on and ride them a couple of times in the round corral, and then we saddle them up with another horse and take them out for a couple rides out in the pasture and just get them used to things. Then I go out on my own.

It's probably about a week before I'm on my own, riding. I usually take a horse for a month. I've been finding it's just not long enough – people expect too much from a month and it's causing trouble, so I think I'm gonna take them for more than a month. It's just not enough time to develop a well-trained horse.

I'm almost done my vet tech course, veterinary technology at Saskatoon. I have six more months left and then I'm done. About half the kids in the veterinary technical course are from a rural background. And that was a surprise because up to this point, they had mostly city kids and very few ranch, very few horse people, so the fact that our class was half and half was a pretty big surprise. Like in the city it's pets, it's companion dogs, little puppies that become "the child" in the family. People are more willing to spend much more on these pets, and that's why small animal vets get paid way more than for large animals – there's just more money comes in for small animals.

I would like to have a better income, but the only reason I want a bigger income is to purchase a ranch. My mom asked me one time why I don't just go in to be a vet and I said, well, it's seven years and I wanted to train horses, I wanted to do it all my life. Actually, my sister and I are planning, when she's done high school, on going down to Texas for three or four months and getting on a cutting horse place where we can work with horses down there, so hopefully it works out.

When we were little we had little toy cows, we had thousands of them, and little horses and little men. We had a barn set up. My dad built some toy barns, so we had one special room where we just had our little ranch, and we had a truck and we'd go to pretend rodeos. When I was in high school, I participated in high school rodeos, where I took part in goat tying, break-away roping, pole bending, and barrel racing. In Little Britches we ended up being in pretty well everything we could be in. It was quite a joke actually. My mom and dad had six kids in Little Britches rodeo at one time. It was for sixteen and under. We were almost all in every event, so it seemed like every second competitor was a Mastad kid. Mom said it was pretty embarrassing out in the stands when some said, "Do you think they're really all from one family? Can't be from one family can they?" But we were.

It never made any difference whether one of the boys or one of the girls did a job. My mom and dad are very strong on that, like if Clay can do it, Myla can do it, like if Myla can, Clete can do it. We do cook, we cook better than the guys, but they do it. And the boys always help with dishes. Supper at Grandma's place at Christmas, the women do the dishes and we're not liking it, but this year was a bit better than last year. We had the boys up, and they cleaned off the dishes and helped us out a little.

I actually love cooking. I love it when I mix up a meal and put it on the table and everyone enjoys it. That's it, but the dishes aren't very high on the list. I don't mind vacuuming, but there are other things I like to do. I don't like being in the house when everyone else is outside. I don't like that at all. Being outside, that's just where I like to be, that's where I'm most at peace.

In the summer of 2001, Jill met Peter Jenkins, who had just bought a ranch south of Glentworth, Saskatchewan. Jill and Peter were married in June 2002. They are now partners with Peter's parents. Jill also worked one or two days a week at a vet clinic. In January 2004 Jill and Peter's first child, Tyson, was born.

PRAIRIE TO PARKLANDS

The parklands, a vast area of prairie interspersed with bluffs of aspen and willow, stretches across central Saskatchewan and Alberta. The southern fringe includes the river valleys of the Qu'Appelle and the South Saskatchewan, while the North Saskatchewan River dominates the northern area. Coulees and valleys along numerous tributaries, such as the Battle River, provide grazing and shelter for wildlife and for livestock.

These parkland areas were the favoured wintering grounds of many bands of Cree who signed treaties with Canada in the last century. Fur traders established posts along the rivers, and eventually the more suitable land was homesteaded and cereal crops were grown.

Today, in the river valleys and on the gently rolling prairie, one can see extensive fields of hay. Unlike ranches on the southern prairies where most grazing land is native prairie, ranches in the parklands are very dependent on hay for grazing and can usually run three or four times as many head of cattle per quarter section.

Women from ranches in widely scattered communities were interviewed. The Catley Ranch is among the oldest ranches in the Qu'Appelle Valley near Craven. The Jahnke family, north of Gouldtown, has ranched in various locations along the South Saskatchewan River. To the north, near Cutknife, lives the Ramsay family. Farther west, in what is known as "The Big Sky Country," the Fentons ranch near Irma. Northwest of Edmonton, at Westlock, the Bibbys primarily raise livestock on hayland.

The ranches in each of the smaller communities are often dependent upon services from a larger centre. Craven is associated with Regina, Gouldtown with Herbert and Swift Current, Cutknife with Battleford, and Irma with Wainright.

DIANE CATLEY

I'm making a transition from lesser to greater involvement on the ranch, next year, when I retire.

Diane (Cottingham) Catley has been ranching with her husband, Irwin, for the past forty years. The Catley ranch was established in 1882, by Irwin's great-grandfather, near Craven, Saskatchewan. Diane worked on the ranch when her boys were pre-schoolers, however, she started working off the ranch when her youngest son started kindergarten. Her "holidays" are often planned to coincide with harvest and other ranch activities.

WE'RE JUST ON THE EDGE of the Qu'Appelle Valley. It's one of the prettiest areas around here because you do get the variants. You have the flat, yet you also get the valley, the trees. Unfortunately, right where our home is located, we don't get any view, but we love the trees because we don't get the wind in the winter and they keep the snow out. Most times we are short of rain, like many areas, but the last few years we've been fortunate. I think we have the best of both worlds, the rural one and the city close by. Regina is just twenty miles away.

DIANE CATLEY

We have three boys. Greg and Mark are quite close in age and then Trent is five years younger – he was to be the girl, somehow the third boy came along! We really enjoyed him. The boys were involved in ranch activities from a very early age because when I was out working, they were out with me and they just fell into it naturally. Then when we went to summer shows, they always came along. I looked at that as our summer holidays. They belonged to 4-H and started showing peewee calves as early as age three.

I actually went back to work, away from the ranch, in 1975. At that time I was doing secretarial work for the Regina Chamber of Commerce. Trent had started kindergarten and our children were in school in Regina, and at that time you had to be six by the first of September to go full-time the next year, unless you'd had kindergarten. He and a neighbour girl both thought they wanted to go, and I thought, well, if I'm going to drive them to school every day, I might as well find some employment in there while I wait. So I found a position from nine 'til noon. That was kind of fun and I had a little extra spending money, too. Then he started in grade one – the kids were gone all day, the bus would pick them up at eight, and they'd be back at four-thirty, quarter to five. That was when they built the Livestock Centre, and they were looking for a secretary, and that was freelance work, so I worked there for a couple of years. Then Dad was diagnosed with lung cancer, and that was really tough – those two years, I spent a lot of time with him. After he passed away, I just had to take a year and get myself back together. I didn't think I'd go back to work. Then I saw this job advertised with the Canadian Cancer Society and decided I'd like to work there – I guess, as part of the healing process. The lady I replaced had been there seventeen years and I wondered how anyone could work anywhere for seventeen years! I did administrative work from 1980 to 1989. Then I was tired of paperwork and a position came up for the fundraising area, and that was more travelling and dealing with people. I've really enjoyed doing that ever since.

Irwin had mixed feelings when I started working. I could see it when I started travelling to Toronto, he found it quiet and he missed my help, too, but the boys were getting older to where they could drive the tractors and do some things outside. I thought that part was good, they needed to be more involved because if I was there, they weren't going to do the work. He was always very supportive – at least, he got used to it, but he's looking forward to my retirement next year. He is getting better with helping with the indoor work. There are still some things that are quite foreign to him, but part of it is that he has more time now. He has become a better hand in the kitchen. Sometimes supper is waiting for me when I get home from work.

I have done pretty well everything on the ranch from time to time. I've checked cows at night at calving time, and checked pastures to see that everything was healthy and in the right location, and I've helped with harvest. My holidays were often planned around harvest, bull sales, or summer shows. I didn't run the combine this year, but I usually help with the trucking or combining. I do chores when everybody's been away to bull sales, so it's pretty varied. Then, of course, I do all the farm books. Before I went to work full-time, I would be outside pretty well most of the day. No matter what the weather, chores had to be done. In fact, when we had that real bad winter, '73/'74, we had to feed the cows – they were all in the meadow, they wouldn't come back up, and the hills all plugged with snow. We couldn't get down, so every day Irwin and I went out with the snowmobile and a sled behind it with little bales on it.

A ranch is a good place to raise children. They get exposed to so much and they have a lot of freedom. Trent spent hours roaming the pastures, snaring gophers. Ranch involvement makes them independent in a lot of ways – I just think that it gives them a real good grounding. When the children were little, we just bundled them up, and we used to have kind of a chop bin in the old barn, and they'd go in there when we did barn chores. When we had square bales they were piled on the wagon – the kids would be on the wagon with them. I know it wasn't the safest, but that's just what you had to do.

The children are involved with decision-making now, but not so much when they were growing up because Irwin's dad was still a partner, so they made the major decisions. We set goals, discuss them – it's been a mutual agreement through the years. Irwin makes most of the ranch decisions, a lot of it's kind of a courtesy thing, sort of like it is in the house when I plan to buy a new fridge or something like that. I've got the say. I think, when it comes to a tractor or something, it's pretty well decided, and then he'll tell me.

When we were first married, it was probably just men who went to the ranch-related meetings. Irwin was very involved with the Hereford Association and he was involved with the Exhibition Board. When the kids were small, really, only one of us could be away at a time, so I was the one who stayed home. I think it's good that women are taking a more active role in these organizations now because they are more partners. I think they bring a different perspective than men do, which is good, for the most part. If they choose, ranch women can be much more involved today than in my mother's time. Just coming from Agribition, there are as many or more women working there than men. I think this is good and I think this will continue. And a lot of women are ranching on their own, which was unheard of when we got married. If I told my dad I wanted to take over, he'd have just said, "You find a husband and then maybe!" I think in the future, ranch women are going to evolve, they'll be equal partners.

When I was growing up our whole social life in winter was the skating rink. And there were always 4-H clubs and dances, and school dances. Through July and August, there were about five girls who always got together, at least every other week, to have a wiener roast or just to be together. The biggest thing in the rural school was the Christmas concert – about the first of December most of the schoolwork was kind of shut down. And being in the primary grades we'd always have recitations and a few songs, but the older kids got to do the plays. We watched them, and I think we knew their parts practically as well as they did! When they put the stage up, that was exciting – we'd run around and do a solo on the stage. Then, on the night of the concert, how exciting it was to peek out the curtains and see everyone there. I think that was probably one of the biggest letdowns when we went to town school, in Milestone, because with all the classes, each room got to maybe sing a song – it just wasn't the same.

Growing up, we didn't go out like girls do now. My sister Betty and I wanted a brother because if you didn't have a date, you just couldn't go places. We had a cousin, Garry, who was the same age as Betty, and his sister was a year younger. There was this small community, Parry, just south, and they always had dances and he often used to have to take us. He'd say, "I'm going to send you three girls to Charm School so I can get you dates, so I don't always have to be dragging you around!"

I started university right out of school because Betty had gone. I didn't know what I wanted to do. The other kids were going, so I said, "I'll go, too!" One of the girls in my class and I were in residence together – we came home for Thanksgiving, then we went back for about three weeks, and we didn't think we liked it. Mother was horrified to think her daughter would quit something after starting it. She had Betty phone me quite a few times. University was great, at first – there was lots of social life – but when it got down to the academics, I didn't think this was for me for another three or four years. So my friend and I quit at the end of October, came to Regina, and enrolled in the Reliance School of Business and got our business degree. Our parents weren't too happy with us, but they helped us find a room.

After marriage our social life changed. It was quiet for the first few years because being close to the city, there really isn't a community right here. Irwin attended a local country school, so never went to Craven. We're in the heart of grain country and most of our neighbours' schedules were much different than ours. They had more free time, especially in winter. Many of our friends, fellow cattle breeders, lived miles away. But it wasn't too long 'til we became very good friends with the neighbours up here. Their children were the same age as ours, and we spent a lot of time with them and gradually became more involved in the community.

I was twenty-one when we had our first son. I wasn't outside much for the first three years of Greg and Mark's lives, more when Trent was young. You can manage with one baby, but not two. Oh, I suppose there's some compromises to make with children – they always come first, so you have to change. We weren't married very long before we had a family, so I hadn't really established a married lifestyle – it's not like I had a career for years and then had children. That wasn't a big adjustment.

Being pregnant limited my ranch activities somewhat because I couldn't do lifting or heavy work, but not to a great extent – I did most of it. Raising babies took a little adjustment, especially with the first one. It wasn't exactly what I thought it would be. They don't just eat and sleep right on schedule. I hadn't been around babies – oh, I liked them and I had done a bit of babysitting, but I didn't realize the constant twenty-four-hour on-call, seven days a week, and there were no babysitters close by. I didn't just have the chance to get away. Every once in a while I took a day off, and I found that really helpful. Being housebound was tough.

I think it's wonderful that my grandchildren are being raised on a ranch. I'm a little concerned – it's unfortunate out there, the financial pressures in the ranching community. The lifestyle is good, but it's getting harder all the time, financially.

Well, we've both lost our parents. It's quite a shock when you realize you're an orphan – it kind of makes you stop and think. Dad, he had his sixty-fifth birthday in hospital when he died, and Mom was seventy-two, so to me that's pretty young. Then Irwin lost his mom real young, she was only fifty-one. That was before we got married – I never got to know her. His dad died at eighty-two. We were fortunate to be nearby our parents.

I haven't had any major health-related problems. I had a bout of skin cancer, which I've struggled with. I'm much better with sunscreen and wearing hats than I used to be. I've had the usual surgeries that come along with age, but nothing else major, and I'm fortunate, I haven't got arthritis, so I'm fairly active. Our leisure time revolves around the ranch. When the weather's right, you have to work. I like to golf in the summer – I'm not a good golfer. I used to curl in the winter, but I haven't been the last few years because once I get home at night, I don't want to drive back to the city. I've taken up quilting – although I haven't mastered it to any great degree, I do like to do that. I sew a bit for the grandchildren and I like to garden in the summer.

The advice I would offer to someone starting out in ranching would be – have a good sense of humour! There's many instances where that's the only thing that saves you, 'cause just so many things are out of your control. You can have the best-laid plans, then the weather will cross you up and you could be devastated in no time. You've got to be flexible. To me, family is everything. Just keep it so that nobody gets really bogged down. Keep things on an even keel, that's what makes it all worth it.

MARILYN JAHNKE

There's never been a woman do this before.

Marilyn Jahnke is a team player in the cattle industry. Her husband, Neil, was president of the Canadian Cattlemen's Association at the same time as Marilyn was president of the Saskatchewan Stock Growers. They ranch north of Gouldtown along the South Saskatchewan River, where they also raised two children, Shane and J.J. Marilyn is a bundle of energy, and with the current BSE crisis, she needs to be. Her wry humour carries her a long way. Though born in the 1940s, Marilyn is very much a woman of the twenty-first century.

WE HAVE A PARTNERSHIP with Neil's brothers. On this place there would be eight hundred head, a little more, and then we have two hired hands here, which really helps when we're on the road all the time. There's no way we'd ever get anything done without them. But we're very fortunate, we've had hired help since we came except for a couple of years. And yes, we've had hired men that don't take direction well from women. If they don't, and I'm the one that's home, then I fire

MARILYN JAHNKE, PRESIDENT,
SASKATCHEWAN STOCK GROWERS, 2003

them. One guy left and then he came back when Neil came home and said I couldn't fire him. Neil asked him, "What are you doing here? I thought my wife fired you?" He said, "She can't do that." "Yes, she can, if I'm not here." This is pretty much a partnership. It doesn't matter if you're washing the cream separator or if you're firing the men, it's a partnership.

We got notification before it was publicly announced that there was a case of BSE in Canada. And it was so hard to believe. We sort of talked about this before it ever happened, but when it happened it was just – it couldn't be, couldn't be. It took a couple of days for it to sink in, and then when it did, my stomach got into a knot and it just stayed that way. Well, the phone never quits ringing. This is really quiet today, but people are phoning from all over Canada and they've all got their own take on this.

Different people react differently and we try to be sort of up for them, but it's getting pretty difficult if this border doesn't open.

They made the announcement in May that we had one case of BSE, and the borders were closed that night or the next morning, that day they were closed, May 20. I mean it was one cow and she never got into the food chain. Just think if there was a ... it's scary. Like right now I don't know what to say, but our system works, they did the trace back, they traced forward, they traced back, they even traced the rendered product. That's fascinating to me – it was announced on Tuesday and before the next Tuesday they had traced all of the rendered product, because she had gone into a rendering plant. Yeah, all the rendering fat went into dog food and they traced the dog food. The system works, like with our ID [identification program]. We needed that after the fiasco in Great Britain and we didn't want an ID program imposed on us by government. We wanted an ID program that we could work with, and that's ear tagging.

If we would have had our ID in place when this cow was born, she would have had her farm of origin tag in her ear and she would have been a lot faster to trace back. They traced all of her progeny really fast. I don't know, it's scary right now – I think it's political right now. The scientific work is done and we did that, I don't know how it's going to come out – one way or the other …

My husband's family has always been involved with Saskatchewan Stock Growers. Actually I think you just get involved and we believe in what the Stock Growers are doing. I like the way it's run and we don't get paid on the Stock Growers. Everything is voluntary and we get very involved in it one way or the other. Like somebody says, the train goes through and you don't realize you are on it until it's too late. See, Marjorie Linthicum pushed me really hard to be a director. Even when I became a director, there were a couple of women that had been directors at one time, but women were not really on the board of the Stock Growers. Then when I was put on the executive, I thought, there's never been a woman do this before.

You think you know something until you get involved, and you find out how little you know by just being active in it. Look at how many times you go to a meeting, any meeting, then you go to a second meeting and you're involved. Somebody asked me, "How does it feel to be the first woman president?" I tell them that we are an equal opportunity organization – the guys make the coffee and I drink it. It's like here, the one that's here is the responsible one, and I think that's the same with the Stock Growers and any of the other cattle organizations across Canada because quite a few of them have women. It comes to a time where you think you have the knowledge so, "Here I am, but ooooh I don't know very much," but I do know where to go to get the answers. That's all you can do, you can't do everything. One person can't.

With me being the president of the Stock Growers and my husband Neil the president of the CCA at the same time, it means we're really hard on vehicles right now because he's going in one direction and I'm going in another. A lot of gas. A lot of gas, in more ways than one.

Sometimes it's really hard to answer and be politically correct. They told me I couldn't use some of my words on the radio, like when they asked, "How do you really feel?" and I wanted to say, "I feel like shit." I couldn't say that. Like they asked what was my first reaction to BSE. "Sick."

Education tax on property has always been my bugbear. I mean the Stock Growers have been fighting that since I don't know how long, but I bet if you went back to all the presidents, they'd all say education tax is one of the biggest problems. We're getting more and more groups involved. It used to be Stock Growers fought it all by themselves. Well, now there's other groups that are getting interested, like SARM and the forage group and a whole bunch of groups are saying, we can't afford to keep paying these taxes. Now we have to come up with an idea – like if we're not going to pay education tax on property, where are we going to get that money and that's a challenge.

We have a committee system in the Stock Growers, and as president I sit on all of them and the land-use committee is just one of them, and a very active one. Some days it's more active than they want it to be – one of the most active committees. We run right on the edge all the time. Besides finance I was also on promotion, policy, and trade at one time or another.

You go in thinking you know everything and you find out you don't know a damn thing. It's all integrated, one thing is tied to another. Like if you don't have the use of your land, you don't have policy and trade, or you don't have anything, like the more you get into it the less you know. You have to know a little about everything, but you don't have to know all of it – you just have to know where to go to get the answers for your questions, that's all.

I've got a really good board, and they're all – I don't think there's any of them that have ever turned me down for a job. I ask and off they go. The land, land-use and promotion, promoting our organization and keeping our membership up. That's really important to a president, I think. And issues like the BSE. We have to be on top of that. Say like when this first started we were having conference calls every morning, so it was like an hour and a half every morning – quite a way to get your day going, to listen to the problems

I think this BSE is the biggest crisis that's ever hit the industry. We were talking to some of the guys about when we had hoof-and-mouth in the fifties, but there again it was a regional thing – they closed the borders to those certain herds, but we were only exporting two percent of our production. Now we export over sixty percent. So if the border doesn't open, we'd have to cut our herds back by sixty percent. We don't want to do that, but it could happen. We're only eating forty percent of our production in Canada. I don't think we can eat our way out of it. We don't have a large enough population. They've done studies since this BSE hit, and we've increased our domestic beef consumption. It's actually increased. There were a couple of days there when it went down, when the news first hit, but now we've actually increased our consumption. One of the media asked me today if there'd be a cutback in producers, and I don't really want to say yes, but I have a horrible feeling the answer is yes, if it doesn't change. It needs to change before the new calf market because right now the feedlots are just trying to get rid of what they have in there. Everything is full – the processing plants, the coolers are all full – and I know some of them have beef stored in reefer trucks.

Actually the local 4-H club had their sale yesterday and they averaged $1.26 for their calves, which is a pretty damn good average, and they were all sold, or most of them, to local businesses. Some of the kids had to take their animals home because there's no place to have them slaughtered right now. Some of the parents thought, oh, no, now we've got to go through all of this again, because you know how kids get attached to their 4-H calves. But the kids took them home and the first time there's an opening they'll take them in because most of the 4-H calves are ready to go. I was a little nervous about the 4-H sale, but I think it's going to be okay.

The compensation package for BSE was put together with input from the industry – like CCA sat on that and they really pushed it. It was to get the compensation money to go into the feedlot industry and the rendering industry, so we can get rid of this backlog that's piling up. And eventually it goes back to the producer, the cow-calf producer. If you can clean out your feedlot, you're not going to let it sit there empty, so you'll bring in some new stuff.

As producers, we can't stand too many good years in a row, so we knew it was going to drop, but we didn't think it was going to drop right out of sight. Always a cycle. Charlie Gracey wrote a book damn near twenty years ago and the cycles are, you're building up and then you get there and you come down – you're always on the uphill or the downhill slope. I mean it's just the way it works.

Inputs just keep going up. I don't think we've ever had a lowered price for fence posts or salt blocks or anything that we need. Those prices always go up and they stay there, but with any commodity there are cycles where it goes up and down – just seems like the down is a lot longer than the up cycle, that's all.

We're here for the long term. Most people are. We might have to really tighten our belts – maybe we'll drive that old pickup and maybe turn off the lights a little faster. I don't know how many more expenses we can cut back on because most people are pretty careful and our margin isn't very big, that's the truth of it. So how can you cut back anymore? We're pretty fortunate – we've got beef to eat.

We used to have two branding days but now we just have one. Any of the culls are taken out ahead of time and the rest are sorted into two fields. We get up and have breakfast between three and four, and the crew are down to the barn at five. They'll ride out and gather the big field, and they'll probably brand a little over five hundred head. And while they're branding those, Neil will take another crew and gather the small field, and by the time he has the small field gathered, the big bunch is branded and they all move over to the small field. They're back in the house by one o'clock. There's sort of a pen – we take them into a corner and we got a little bit of a corral there for the three sides, and they usually block the other off with trucks. And then we rope them. And wrestle.

I like to have any of the baking for branding done before the convention. The Stock Growers is Sunday, Monday, Tuesday, then we have Thursday, Friday to get ready, and we brand on Saturday. But I like to have all my baking done before Stock Growers. And then all I have to do is the cooking. The numbers can vary – sixty, seventy people. I don't get to ride at our branding. I have to cook. Four of us move all the cows in the spring, but forty gather them at branding. Yes, we have prairie oysters. Yes, and lots of beef – it's all beef, even the prairie oysters are beef.

One year later Marilyn stated, "Just when we think we're getting somewhere with this BSE, *then we're not. It's been a real roller coaster, but I don't know anyone yet who can say they've lost their place because of* BSE. *We'll be around, and maybe we'll be an even stronger industry."*

MARILYN RAMSAY

Except for my morning coffee, after twenty-three years of marriage we have not had two seasons, let alone days, which you can say are the norm.

At the Maple Creek Purebred Breeders' Annual Bull Show and Sale in March, 2003, a horned Hereford bull contributed by Carl-rams Hereford Farm of Cutknife, Saskatchewan was declared the Grand Champion. The Ramsays, Cal and Marilyn (Peterson), along with children Carl and Robin, work together on the farm/ranch operation six miles north and two miles west of Cutknife. Marilyn said it was a thrill to win at Maple Creek: "Everybody's competing but you still pull for each other." This year Marilyn, born in the late fifties, was nominated for and is presently serving as a director for the Saskatchewan Hereford Association.

I GREW UP on a grain farm at Sceptre and I've always enjoyed the cattle, that end of it, the outdoors. I went to elementary school in Sceptre, high school in Leader, then I'm a graduate from voc ag at the University of Saskatchewan. I worked for Garry Dorchester two seasons travelling with the family, taking care of the wagon, horses, and grooming. I wanted to learn to ski, so got a job at Marmot

MARILYN RAMSAY TEAM PENNING AT THE SASKATCHEWAN FINALS, 2003

Basin at Jasper and worked as a lift operator for a winter, and I also went to New Zealand with IAEA [International Agriculture Exchange Association] and worked on a sheep ranch.

The university vocational agriculture program is a two-year diploma course, and that's where I met my husband. Out of the two years, 106 of us graduated. Four were female. Now over half, I bet sixty percent, of the graduates from agriculture are female. So it's changed. Girls are getting more into the role.

I worked on the grain farm at the university for a couple of years, and then it was marriage, home to the ranch, I don't work off – I'm a full-time ranch woman. The farm/ranch has always been a purebred farm, purebred horned Herefords.

I don't think there is a typical day, so whatever has to be done ... everybody just chips in and does their thing. The cattle come first – right now we're into calving and the cold weather was not a pleasant thing – three weeks of forty below weather calving cows. Every calf was under the lights. We calved right through it, so lots of walks to the barn. We share, but I often will do the nights. I'm good until about nine o'clock in the evening, and then I have a few hours sleep and then I'm good again from three on. And Cal, he likes getting up in the morning, so I think we'll start more the beginning of February again. It's hard to justify the early calving. The bull we brought to Maple Creek is an April baby and he's the champion, so there you go.

My main thing would be calving, the bookwork, "gofer," and the combine is mine in the fall. I learned quite a while ago that it's far easier to drive the combine than the truck. We have a grain operation, too. We farm about twenty-five hundred acres. We use a semi for hauling grain at home – you have to learn how to drive everything and I'm not legal without a license, so I took my class one driver's license. When I was taking my air brakes I was the only female in the room, and our instructor had his muscle shirt and the crude language and the whole bit. But I happened to notice that when I went back probably within a year to stop in at the office, that the instructor was wearing a suit and there were about four girls taking the course. If you drive up and down the road now you'll find a lot of women behind wheels of big trucks, so it's changing. I drive a lot and I see in the future any stock trailers will need air brakes to be safe on the road.

Motherhood scared me. I think it's easier to raise a puppy. And they don't send manuals with the kids. Everyone's different. We have two children. Carl's twenty-two and working on the rigs, coming home when he can, and Robin's at Olds College taking ag business. They both say they're coming home. We say they have to leave for at least two years. They're four years apart and they are extremely close. They both love the cattle. The cattle will be – somehow, somewhere they will make a living off the cattle. They both always have a "hands on," and you can't learn any younger than right now.

Carl has always been protective of his baby sister. When they were little, let's see Carl was four – I artificially inseminate the cows, and the year when Robin was a baby, her time schedule would be made up around inseminating the cows. Carl being an early bird was up. Robin would get up, we'd get her set, get her back to sleep, have his lunch packed, and out to the corrals we'd go and we'd have our cows ready. So we're at six o'clock in the morning now, and by the time I would get a cow in, Robin was sleeping in the truck. Carl would be in our little shack and he would know what instruments I would need on which kind of cow we were doing. He'd have it set up and he'd watch his little sister – he was four. I think now, how did that ever happen?

If Carl were to be married right now, he would still have to have an off-farm job to support a family. Our farm is a fair size and it should be self-sufficient. I don't like it when agriculture can't be self-sufficient and you have to work off farm just to have a quality of life. It should be a business, not a lifestyle.

I would not want to raise my kids in an urban setting. I couldn't. Like to take the garbage out would be enough responsibility for them. They've learned responsibility from a very early age, which is in life-and-death situations if they're going to check a cow, they go to check a cow, and they've learned to drive at a very young age. We're nine miles from a hospital. If something should happen, they don't want to be helpless. I think they have to learn. I think they have to grow up and mature and be responsible more on a farm than they do in town, maybe in a different way.

I love dogs. I'm a very elementary trainer but if everybody leaves us alone, my dog and I can get a job done. There was one day – wives should learn not to laugh at their husbands. I had a pup and of course my little collie pup has to come with me all over. It was one of those bad spring days where the corral looks more like lakes and islands, and we had two cows deciding to calve on their own private islands with my pup. One needed assistance and one wasn't really friendly, so you shouldn't laugh out loud when your husband is trying to help the one cow on that island and the cow on the other island decides to chase him, and then the cow he's helping chases the pup. Both calves came out fine – husband extremely wet and red-faced – and the pup lived on a little island all on her own for about a day.

Another time my dog was a small Heinz 57. She went everywhere with me, and I was going to be the good farm wife and raise my own chickens, turkeys – to save money, right? Well, Cal really is not "dog people" and one day just wanted to go for a Sunday drive without my four-legged friend. Bad move. We came home to a very content dog lying on our step, surrounded by thirty dead birds she had killed and brought to the house for my inspection. Solved that problem real quick. From that day on *my* dog always travelled with me – she learned to like the kids. Her and Cal never really saw eye to eye, and I did not like raising chickens that much anyway, so never did again.

Cal has since learned to like dogs. If it had not been for our German Shepherd, Carl, our son, would probably have suffocated in a badger hole he had crawled into when he was a very bad two-year-old.

As for farm accident things, Carl was kicked in the face by a colt one time, which was extremely scary. The colt was just playing and kicking up her heels and he happened to have his face in the wrong spot. So it wasn't a strike, but it was scary.

Once both of the kids were driving the John Deere Gator that they used to come out with, the one-seat type. Well, they took a pop out to their dad in the field – Robin would have been four, so Carl would have been eight. Robin was sitting in the back – it's got a little box, a carry-all box type of thing – and on the way back from delivering their pop, they turned too sharp and one of the tires was flat and they rolled. Carl got thrown free, but Robin hung onto the box and was completely pinned underneath. Luckily her dad saw it and was able to run and overturn it, but she had inhaled some loose mineral, cattle mineral in the bottom of the box, and it burned her throat extremely badly. So she spent the month in intensive care over that one. They kept a tube down her throat so it wouldn't close, for the month, and then she healed on her own, but surgery was definitely a possibility. And the doctor rode with me all the way to Saskatoon in the ambulance, which is scary – him sitting there with his trocar in his hand in case she decided to close off her breathing. We know we're high risk, and we have to be prepared and handle it properly. The first two hours of any accident is a crucial time, so we have to be prepared and keep our heads.

Growing up we rode a lot of horseback. If I wanted to go visit my friends, I'd have to ride. I've taken a colt and taken him right on and it is very rewarding for sure, but I think it's a young person's job. I do send mine out now if I want to get them done right.

One time – there's five women – we started an afternoon ride in the old community pasture and we weren't watching the weather, we weren't watching where we were going and our map reader got us lost. The five of us spent the night in a very electric rainstorm out under the trees. And the pasture was huge. Our husbands – only one husband was really worried about us. The other ones said, "Well, they'll get home. What are we going to do anyway?" They didn't know where we were. I'd gone for a very nice afternoon ride. I had absolutely nothing with me. Thank goodness one of the ladies smoked because she could get a fire going. They put me beside this fire with my saddle blanket over me because I was as close to hypothermia as you could get. I was very wet and very cold, but ever since then when we've gone for our reunion ride, we've gone prepared. My attitude towards smokers changed after that night.

Actually, on one of our rides, a mule caused my worst wreck. If anyone offers you a mule named Amos, don't take it. Mules are totally different than horses. This one year I was going to be really prepared, so I took my pack mule and he irritated my horse so bad that on the second day he got his lead shank under the tail of my horse and my horse decided to buck, and there I was, lying in the middle of this community pasture flat on my back, and when I woke up the rest of the women were standing there looking over me. Apparently I'd been knocked out for about ten minutes and they thought I was gone.

They didn't want to touch me. They were wondering, where can we get help? I was lucky. I think the wind was knocked out of me more than anything. It could have been very scary because we were quite a ways from nowhere.

One time I was riding, visiting with a gentleman, Vic Oddan, and he was driving his covered wagon. We were on a little bit of a weekend trek and I was approximately six feet away from his wagon in the mud, leading one of the kid's ponies, and my horse almost disappeared out of sight. We got out of it, but it was quite scary. I was six feet away from solid ground and it was just a small area, probably an eight-foot circle but I was unlucky enough to find it. Actually the momentum of everything – my horse – we all came out together except my camera. It's still in the mud.

To be very truthful, except for my morning coffee, after twenty-three years of marriage we have not had two seasons, let alone days, which you can say are the norm. A rancher's wife better be able to change her plans in a big hurry or she will go nuts. I am very fortunate I have the physical ability to do some of the things I do, but also I enjoy it and would not have it any other way. There's always interesting things happening and if you couldn't laugh at them, it would be a long day. Definitely keep your sense of humour and I think your health goes right along with it.

I have learned a lot about cows, decided the stereotype farm wife was not my cup of tea. Cal has learned to do dishes. We work together, we play together. Oh, it's had its moments. As long as he remembers, yeah, I really am the boss, we get along fine!

We have a pencil drawing in our kitchen with the caption:

GOOD HANDS MAKE A TEAM. ... A GOOD TEAM MAKES A FARM.

That works for us. If you can enjoy your work, it's not really work. It's just a bonus.

My friends are people of all ages, sexes, religions, nationalities, and I have some strange ones, but the common denominator is probably love for dogs, horses, or *cows*. I have closer male friends than female friends, just because I don't talk about knitting and cooking very often.

We strongly rely on marketing our cattle from our yard. Maple Creek is one of the only sales we come to but we also have to rely on return customers, so you have to keep ahead of them, which is harder ... so it keeps you improving. I think the younger ranchers are far more into the genetics, the numbers. It's a different game than just going and looking at an animal you like, and we have some young ladies that are right in there trying to start.

Town life's not for me. I see myself retiring with a cabin in the north forty somewhere, or just far enough away that I can come home and help if I want to.

BSE sucks. It has shown us just how small and vulnerable we really are. Ranching and marketing our animals as we have done in the past will change. If we want to stay in the cattle industry, we have to let the past be the past. To survive we have to have a positive outlook towards the future and find a new way to make it prosperous.

On a personal note, it was my hope this spring that our twenty-three-year-old son would come home from his job in the oil field and expand on the ranching side of our operation. His mother did not raise any dummy – he will not be giving up that oil cheque any time soon. But on the flip side, our daughter at nineteen has not had a taste of big off-farm wages, still wants to be part of things, so life will go on.

I think Robin's age group will have things easier because nothing is really expected of them – no set female stereotype to fall into, the doors are wide open. I think we are seeing a lot more ranch women in the making, not just rancher's wives.

ROBIN RAMSAY THROWING A LOOP ON A CALF, 2003

ROBIN RAMSAY

If I'm an old spinster with my forty cats on the porch, I'm going to be a rancher.

Robin Ramsay is the daughter of Cal and Marilyn Ramsay of Cutknife, Saskatchewan. Robin graduated from high school in 2002 and is presently attending Olds College in Alberta, where she is enrolled in the two-year agriculture business program. She has a "hands-on" involvement in the ranch at home, along with her parents and her brother, Carl.

EVERYBODY HAS their role at home. We all get along super. Mom does most of the work. No, everybody pitches in and if they didn't, they probably wouldn't be there. Without one, the whole thing wouldn't work.

This year I'm at Olds taking ag business. I miss home but it's good. Now I'm rodeoing quite a bit and going to all these bull sales and Junior Hereford meetings and things like that. So I haven't been home – I was home once since Christmas and that was last weekend – I needed to go home.

I'm majoring in finance because I know nothing about the book side of things. That's Mom's department at home. So I'm trying to learn that.

I team rope and break-away rope and college rodeo. I head and I've been team roping since I was about thirteen. It's fun showing up the boys sometimes. I get invited to a few different brandings, which is nice because that's a huge honour – like that's neat. Those are my favourite times of the year.

In the spring and fall I work for different people at all the bull sales like Calgary and Regina. I just go when people hire me. It's nice – I get to see their cattle and meet all these people on a bit more friendly basis. They're not just acquaintances anymore.

In 4-H we were always in charge of doing everything and that was the only way to learn. Mom and Dad wanted us to learn ourselves, and then we did more and more at home, and my brother – he does a lot better job clipping than I do but he's helped me quite a bit. And then we go to a few clinics here and there, but it's usually just practice. I was in 4-H beef and light horse and draft horse driving and rodeo, and dogs for a little while – always beef, though.

I have a huge mix of friends. I have more male friends than female friends. I don't know why, but my favourites are the ranch kids. We have so much more in common and so many similar ideas about things. If I'm an old spinster with my forty cats on the porch, I'm going to be a rancher. And I think it's the only place to raise kids. Guaranteed. I couldn't think of a different life for myself. It would be my ideal life. I wouldn't want to be a rock star. I don't think I would raise kids in the city. There's a huge difference in the kids, in the behaviour, and here you have responsibility growing up. You have, like, part ownership, something to look forward to. I don't know, it's kind of scary nowadays how things are going – cities are getting bigger and more drugs and whatever else.

I see a lot of my friends who have it pretty bad and divorce and things like that. I don't know. I just liked how I was brought up and where I was – it was pretty fun.

I have a really strong interest in counselling or social work, and if I *had* to live in the city and I know to do that kind of a job – you'd have to be in an urban centre, otherwise you wouldn't have anybody to come see you. But I would do that, I would love to do that someday. Maybe I will. It would be nice to see if I could make it work.

Dad breaks teams in the wintertime. We feed the cows with the team. That's his hobby. He really enjoys that. When I was about nine years old – I might have been younger than that – my Dad was out seeding and I wanted to show him I could drive the team, so I went and brought in the draft horses and harnessed them up just to a little buggy. I had to use a stool and a pail to get everything on, and anyhow I took them out to see Dad – to take him his lunch with this team, and he was all proud of me. I tied up the horses to the willows and then went and talked to him in the tractor. When I got out of the tractor, this horse was like running halfway down the field and the buggy was bouncing upside down and around and I had to walk all the way home, but I did it.

Things are pretty comical at home. Dad hardly ever gets upset – you think he might, and he either laughs it off or says, "Well, it happens, you know," and he tells company, "It gets a little western around here sometimes," and everyone laughs. Every day there's something different that happens. You know, a cow will hit the end of a fifty-foot rope and almost wipe out the whole family or anything.

All four of us are involved in the family decisions, like whether to leave the horns on the cows or not. We have a family vote and that's it. Dad gives quite a few management roles, like to us kids, too, so I think it's equal. We have coffee table discussion – they call it a family meeting, but it's over dinner and we all know it's coming and we tell each other how we feel and make a decision. We usually talk everybody into the same thing anyhow.

I think since we've been growing up, Dad and Mom have been trying to make it so we could get along with each other so that we could work together. I know they've told me, like, because I was pretty upset when I graduated. I thought, I'm going to be a guest in my own home, you know, like, I'm out. It's not that big of a place to handle four families and you never know when you're even going to get married or not, but they told me – like me and Mom went out for a ride on the hillside just about my last day, and she said, "You're welcome here, you're welcome to come back, there's a spot for you. We want you to come back but it's totally up to you." So, that was nice – like, I know that they want us to come back. Mom said, "Well, we'll raise chickens if we have to, to make it work." Like because the land base isn't big enough for four families, but she said, "You're welcome to come back just as much as Carl."

We have a really strong relationship with the Hutterite colony about three miles from home, and those are like Mom's best friends, and they're our best business partners as well. They buy all their bulls from us. Some people don't understand the friendship we have with them but they're the nicest people around. Like, they built a dairy barn and the whole thing was tiled, and they brought over a whole bunch of tiles and a tiler one day and said, "Marilyn, do you want your bathrooms done? We got extra, you know." They tiled the walls and the floors and everything for Mom. Right now, for Dad to come here, to this sale, somebody had to go check the cows and whatever, so a few of the guys will come over and help out. They're awesome friends.

I think ranching gives a person quite a personality, just because your days are filled with such that you never know what's going to happen. To sit down with a rancher and visit, it's pretty funny. Everybody gets along so good, probably because they haven't seen anybody in awhile. It's the character and the people that make people want to still do it. Like to come to this bull sale – these people are all great friends and to get together once a year is a big event. It's a big social event more than anything. Everybody teases each other and you walk in and everybody shakes your hand four times a day and gives you a hug and bugs you. I consider all those people awesome friends, and if they ever needed anything, we would be there for them, and we – if we ever needed anything, they would be there for us.

I think ranching is the one thing that's going to stay more the same. Like, it's because of the lifestyle and that's what people want, is to have the lifestyle – to get up, watch a sunset, take time for family and friends – and I think it's the thing that's not going forty million miles an hour. You know, that's why people are in it – because it's not getting faster and bigger – and you know, the money is still not bad and people are still surviving, so I think it's not going to change a whole lot because people want to go backwards more than forwards. I think it's more of a race for the dollar in other places, and in this everybody is broke anyhow, so you just got to be friends.

Like ranchers, they all say they wish they were born a hundred years ago, you know. So that's the kind of life that they try to make, and they can do that when they're out in the hills by themselves.

DORIS FENTON, IRMA PARADE, 2002

DORIS FENTON

I like living alone. I don't have to cook for anybody.
I eat when I'm hungry, I drink when I'm dry,
and I don't have to work anymore.

Doris Fenton was born and raised on the farm where her father had filed on a homestead in 1905. Her father, Walter Gray, the youngest of seventeen children, had come from Tennessee, and he returned to that state in early 1914 to marry Doris's mother, Lily May Cherry, and bring her to his new home. After teaching at rural schools in the area for several years, Doris married Stuart Fenton of Irma. They raised a family of one girl and four boys while operating a large farm and commercial/purebred Hereford ranch. Today Doris lives on the place near Irma where she and Stuart started out more than sixty years ago. She still rides when she can and takes a great interest in the ranches and lives of her children and grandchildren. [Judy Fenton is Doris's daughter-in-law.]

WHEN MY FATHER came from Tennessee, he found a quarter he wanted – it had already been filed on, but he stuck around all summer and finally the man gave it up. So he filed on that homestead. He and whoever was living with him built a little shack of tarpaper and shiplap. It was a terrible winter, 1905 and 1906, which old-timers still talk about. The snow was three feet on the level, which it rarely gets in this part. They hauled logs from the Battle River with a team of oxen, and the nearest place was about three miles. It took them three days to make a road so the oxen could get to the river. After that they had to chop the logs and bring them back. Then they built a log house, the one that I lived in until I was eight years old. They got their mail and their groceries from Vermilion, which was thirty-two miles. My dad made himself a pair of homemade skis out of floor plank, and he went on skis across Buffalo Coulee, across Grizzly Bear Coulee to Vermilion. And part of that winter he helped dig the basement for the New Brunswick hotel, which is still used in Vermilion. And he put up ice for the hotel on the Vermilion River – it must have been a little bit deeper than it is now. I spent a year going to high school in Vermilion and I used to swim in the Vermilion River – it wasn't much of a stream. Also it was quite sluggish because we were green when we came out from the algae or whatever that was.

Mother was scared of horses, she was scared of cattle, and she was scared of chickens, but she pumped water by hand for the steers, great longhorn things that my dad would buy and feed all winter. She milked cows – she was deathly afraid of them, but she milked cows, and she raised about three hundred chicks every year under setting hens, even though she was afraid of them. But she wouldn't ride horses. She had ridden horseback to school when she was a teacher in Tennessee, but she was afraid of all our horses. Except there was one horse that my dad broke as a two-year-old, and my mother decided that was going to be her horse and she wasn't afraid of it. But it had more bad habits that any horse I was ever acquainted with. Not dangerous habits, but really bad. If you said anything that sounded like "whoa" she would stop dead – then my sister would often fall out of the buggy because she would stop so quickly. She would shy and throw us off because we weren't allowed to have a saddle 'til we were about twelve or thirteen years old. If you hit her with a switch, she would kick out and sometimes break the dashboard or the harness. She never bucked, but that's about the only thing she didn't do. But mother was not afraid of her and she would drive her, but that was the only one she would drive.

Mother always tried to make us suspicious of men. We always had hired men around. She wouldn't allow us to go to the bunkhouse, and we weren't allowed to help with castrations and brandings when I was a little kid. We knew all about what went on, but we weren't allowed to be out there. Mother was raised in the southern States where there were certain things that ladies did not do. Mother's mother died when she was very young, and Mother was raised by a black woman. Mother had different ideas about black people than what is considered politically correct now. She wouldn't sit down and eat with a black person – there were lots of things she wouldn't do, but she said, "I guess I loved a black person more than anybody I ever loved in my life." And that was the lady who raised her.

Of course, it was considered a terrible thing to have a child out of wedlock, and of course, all the old biddies talked about when a girl "had to get married." I remember saying to my mother-in-law when we had been married a year, that they couldn't say I had to get married. Grandma had a different way of saying it – she said, "None of my children have ever disgraced me." And that is what she meant. I know of lots of women who were pregnant at the time they got married. But I wasn't counting – I didn't have to, all the old ladies counted. Mother never talked about it much. Except for my dad, she never trusted anybody much. At least not any men.

I was the oldest of two girls, so I was Dad's boy. I went with him everywhere, and he would send me to get cattle or find cattle or horses and I wasn't allowed to come home until I got them. He might be sitting on a hill somewhere to see that I was all right, but he wouldn't let me know that. So after I was married and I would be out doing the same things I did as before, I would be hunting or I would be moving cattle on my own, and my kids would say to their dad, "When is Mom coming home?" And he would say, "When she gets finished." And it might be pretty late, but that was just the way I was brought up.

My sister could ride too – she helped me a lot with the cattle and what not, but she was a real good cook and she liked cooking. I was a year and a half older than she, but nowadays because of her diets and her careful makeup and her nice clothes, she looks at least twenty years younger than I do. She never allows herself to get tanned anymore. She was a tomboy, too. There weren't any of the boys at school who could beat up on us because we practiced on each other.

We rode or drove to school. It was only two and a half miles. It wasn't a big school when we first started, but when I taught there years later, there were thirty-six students all in one room. It was called Battle Heights School. I went there until I was finished grade nine, and the teacher had so many students that he couldn't give me any school time, but he stayed after school and taught me two subjects each night. Then I went to school in Wainwright.

Our crop was hailed out in 1929 and so Dad didn't have any harvest to take off. I spent the last part of the summer herding cattle on the hailed crop, and we could only let them in for a little while at a time at first. I was breaking my first horse at that time and I really got lots of riding in because the cows got so tired of being chased in and out of that place, they were just furious and I had an awful time with them. Anyhow, it was good for a horse and she learned a lot of things.

All the business part of Wainwright had burned down that summer and they were still rebuilding. At that time many of the merchants had their homes upstairs over their businesses, so there was no place to live and every house was full and bursting with people. So my dad built a new house in Wainwright. He thought that was a good thing to do and he could sell it when we were done with it. We lived in a tent while he was building it because he didn't get it finished in time for the start of school. September, October, and part of November we lived in that tent with nothing but an oil heater to cook on and to keep us warm. It snowed and there was two inches of snow on the outside of that tent, and we were so homesick for the farm. We had a friend staying with us who had come in from the country too, and she was so homesick that she had to go home, but Mother would not allow us to go home. It was really a terrible time for us. When we would complain and want to quit school, Mother would say, "That's fine if that's what you want, I'm sure the Chinaman will take you on at the restaurant." There was nothing I hated more than washing dishes and my sister wasn't very taken with washing dishes either. So we stayed in school. Then we moved into the house and Mother stayed with us all winter until spring, and she went back out to the farm. Dad had to be out there to look after his cattle in the wintertime. Then the second year we also batched at the house, but by then we had lots of friends in town. Most of them came from the country as well, so we weren't so homesick. My sister was taking piano lessons – we had the piano in town and we had lots of fun.

There was no grade twelve in Wainwright, so I went to Vermilion for a year. I was used to being away from the farm and I had good friends there and a nice place to board, so I wasn't homesick. I learned to do a lot of things I never learned to do in Wainwright. We got to skate more, and I had a roommate who was from Kitscoty. There were three high school students living in that house, so I had lots of fun. And I'm still friends with them.

We were required to go to normal school for nine months before teaching. Camrose was still open at that time, and that was a very good year, lots of new friends – a few of them are still around and I still see them at times.

I taught for nine years, 'til two years after I was married. After that I subbed some. I lived with my folks after I was married when I was teaching because I had a school that was about four miles away, that was Paschendale School. Then I would ride on weekends – I would go home to the farm at Irma. And then when winter came, I would ride back on the saddle horse. I had brought my own horse with me. I kept her with me all the years until she died, that was the first horse that I broke.

I didn't meet Stuart until I was in my twenties although we had only lived twelve miles apart all our lives. It was 1941, but Stu didn't go to war because his parents were elderly and he was the only one to look after them. We lived with Stuart's parents for two years, and then we built a little shack beside them. And it was meant to be a granary later on, so I couldn't have the windows where I wanted them because we had to put the windows where the grain spout would be. It was down where the calf shed is now, and we lived in that until Carl went to school, five or six years anyway. I had three by then, Carl and Barb and Henry. Then Stuart's dad got sick and went to Edmonton and had cancer surgery. They came back to the farm after Grandpa got out of the hospital, and then they decided to move to Wainwright and Stuart and I moved into the house.

We started out in the cattle business on our own by buying other people's pail bunters in the fall. Stuart had a friend who was retired and was quite astute about cattle. He would go around picking up a few pail bunters here and there that he thought we could afford. The first steers we got cost us twenty dollars – we got eight head for twenty dollars a head. And then he found some more for us and they were better, they only cost nineteen dollars. Stuart's brother-in-law, who lived closer to the river than we did, had some land close to the river, and he was having trouble with his neighbours who ran a lot of cattle and never fixed fence. So he said if we would fence that land between his crop and the river, we could have the pasture for free to run these pail bunters. We bought them in the fall and we wintered them on hay that we put up on the army camp.

We started our cattle herd with those pail bunters and we kept on, and finally, I think it was about 1946 that we branded our first purebred Hereford heifer. We bought three registered heifers from a neighbour, and then we bought ten cows later. Of the first three heifers, one was a shy breeder and one had poor calves, so we should have been discouraged right from the beginning. We already had some commercial cows and we were getting a few more all the time, and we had decided that with the land we had, we should have cattle and not be farming so much. Of course, we kept on buying farm land too, more and more.

There was a buffalo park called Wainwright Buffalo Reserve, and when the buffalo got so numerous that they didn't have enough feed for the winter, the government bought a bunch of land between the park pasture and Ribstone Creek that had a lot of hay on it. In 1938 word was received from Ottawa that the Wainwright Buffalo Reserve was to be cleared of all animals to become a national defence area. Sam Purcell slaughtered the buffalo for the government. The roundup continued all win-ter, and by the spring of 1940, the buffalo park became deserted rangeland void of all animals. That's when they started the Army Reserve there. The younger animals went to Elk Island Park. After that we put up hay on that ground.

I did pretty much the same work after the children were born as before, but not in the wintertime. My sister and her husband had sold their farm and got an auto court in Penticton, and so I got two saddle horses from them – one was kind of a bad actor, but I kept the other one, a little stallion, and I rode him for years. I was breaking him after Barb was born. Grandma had a girl here helping her in the wintertime – I think Grandpa was sick by then – and this girl had a bedroom upstairs in their house. And she looked out across and saw me get bucked off way over there. Stuart was miserable – he always rode the old horse and I rode the young horse snubbed up to his. Whenever the young horse would start to act up, he'd let them have a little more rope and a little more rope, and this time he wasn't thinking that I didn't have strength in my legs yet because Barb was only a month old. Anyhow, there was a lot of snow and it didn't hurt me at all, and he turned out to be a good horse. I did quite a lot of the same work, and of course, we didn't have a lot of cows at that time, but I remember when Carl was first born and I was in the house and Stuart was hauling straw and I couldn't go. I was watching out the window in the kitchen, bawling because I couldn't be on the load of straw with him.

Stuart learned to help in the house; he had to. I always made him cook the steaks so I wouldn't have to listen to the guff about how they weren't done right. Stuart shared in parenting to an extent, but as little as he could get by with, though he'd play with the kids in the evening and so on, but he wouldn't change diapers if he could get out of it. The hired men that we had here would always watch the kids for me after supper. We had old John – he came to us when he was in his eighties and he lived to be ninety-three, but he was in the Wainwright Lodge in the last years. He would watch the kids at night and he would stay with them, and I could go to dances and things because he was very trustworthy.

The kids had to get out in the yard and they had to chase cows, but it seemed to me like they were always in the wrong place. Gradually they got better and they all took turns riding with me. I didn't let them ride with saddles when they were little because that's the way I had been brought up. They were pretty good riders, pretty good with the horses. They were good with the cattle too. Whenever the seed started to go in the ground in the spring, the cows were mine. I had to sort out the cows and calves and put them out with the bulls. I had to dehorn the commercial calves and do horn weights and all that kind of thing. My son Jay and Jackie, his wife, live here, and poor Jackie! In lots of ways Jay is like his dad – when it comes spring and it's time to work on the land, the cows are Jackie's. Jackie was a town girl. She can ride, but Jay expects her to do everything with the cows once seeding starts. So she does all the things I did. I branded and I tattooed, but I never did castrate. We had a commercial herd, as well as the purebreds, until about 1974.

I treated Barb a little bit differently than the boys because she was a little easier to deal with. They didn't think they should do any housework because Dad didn't do any, but they had to do some. I wanted them to enjoy school, but their dad used to keep them out of school a lot to help, especially the boys. Jay used to keep track of how many days they missed school, and if it had been all in one chunk, it would have been two months of school. Of course, as far as sports went, they got all the encouragement in the world from their dad – he hauled them for miles to play ball or hockey. 'Course, when he wasn't hauling them, I was.

When it came to making decisions, I always had my "say so." We didn't always do what I said, but very often we did. In 1979 when we bought the ranch down south near Del Bonita, I was left here. I had the two younger kids, Jay and Alan, and we got along fine. I guess Henry was here when Stuart first left for the south, but Stu was gone for eight years. He came home the odd time. So basically this place was up to me and the two boys. They had always worked and helped at home a lot, and they helped with making the decisions. We always talked about things and we still do a little bit. I always tell them if I don't agree, but I don't make any fuss about it because they have to learn to make their own mistakes. They may learn a little bit by hearing me tell about my mistakes, but they really don't believe it until they make their own. About three years ago, we had a real bad year here, hardly any rain at all. Then they asked if it was this bad in the 1930s. And I told them, "Yes, if we had been farming the way we were then, it would be just as bad. Still using horses, and by the time we had one side of a field ploughed, the other side was too dry to plant. And your implements took so long to do everything, and these new things do a lot better job of farming than in the 1930s."

I really much preferred being outside always. Now, I don't go out as much. I try to keep my big mouth shut and my big nose out of things. I do help Jackie once in a while, I do ride some, but I don't have my own horse anymore. I can't lift my saddle on the horse anymore, so she saddles up my horse. She has enough to do without babying me, so I don't go as much as I used to. I run a lot of errands, I drive back and forth. I sometimes make two long trips a day, depending on what's needed and where it can be found. But I try not to mix in. There was a time when I didn't think branding could go on without me, but Jackie's been out there branding ever since I've been home from B.C. I have only got out once, and all I did was clean the ears for tattooing. She's afraid I'm going to get hurt, she's afraid for me to put poles behind them in the chute. I don't think I'll get hurt – I've done all these things all my life and I've only ever been run over by a cow once or twice. I've got good bones. I had a bone test a year ago and my bones are good. I never broke any bones until I broke my collarbone when I was past eighty. Lots of people break collarbones.

If I can be on horseback, that is the best. Last fall Barbara came home and she only had a day. Al and Barbara and I rode and checked all the gates and fences. I don't know how long it had been since the three of us rode together. It was a real nice day. I ride with Al and his kids every once in a while. I think it's only been about four years now since I haven't helped Al put out his herd bulls in the spring.

I suppose I had some pension coming from teaching, but I have thought that since I had a husband and I own some land that I shouldn't be taking a pension. So I left all the money in the pension fund that I had paid into when I was teaching. My salary was so small that I don't think it amounted to much anyway. So I have no pension. So all my money came from here. Nobody gets rich ranching or farming, especially when it's like the men in my family who want to buy more land and more land and more land. It didn't just stop with Stuart. Henry and Al and Jay are all the same way, more land and more land.

I always wanted to be a mother but not now, never now, just later. And along came five of them. And with the last two I thought I would never make it. I was old when they were born. I got a late start but I had a late finish too. I was twenty-nine when Carl was born. Jay was born on my forty-second birthday and Al was born a year and a half after that. My first ones were all two years apart, but that just happened, and then nine years later I had these other two.

I still rode when I was pregnant. My first three children were delivered in Wainwright by the pioneer doctor who delivered me. When I went to him I told him I had been riding and he, being a pioneer, thought nothing of it – he said just not to ride so fast and just carry half a pail of water and half an armful of wood. I told him I'd been riding because I was riding the horse behind the harrows – I was riding in the river and carrying on with my usual things. For the last two kids I had Joe Kallal's palomino, and he was so smooth to ride, he was really a nice horse. You could catch him only on the left side and he didn't want to be petted, but once you got on his back he never did anything bad. He was such a good horse. An old lady that I boarded with once said her grandmother had been a great horsewoman, and she said she thought that people who ride horses a lot have difficult births. The wrong muscles are built up – the wrong muscles are strong. I had long labours, it was three days for most of the kids. I was in good shape too, because I was active. It took a long time for the strength in my legs to come back.

Stuart wasn't there when the children were born – he got out as soon as he could. With my first three, I was kept in the hospital for two weeks. I'm deathly scared of babies. I have a new great-granddaughter, Jennifer's baby is a year old, but I only hold her when they plop her in my lap, I'm afraid. I wasn't afraid once I had my kids and got working with them, but the first one was big and strong and he stood all kinds of experiments. He weighed nine pounds and nine ounces when he was born, and he was not fat – he was just big. He is still the biggest of all my children. He is six foot three and his son is six foot seven. I had Auntie up the road, the kids all called her Auntie – Stuart's brother's wife – so I did too, and she was a good friend and she was good with kids. She was relatively uneducated but she had lots of practical knowledge, and she babysat a lot for me when I just wanted to go for a short time. I never left them with her for a long time because it wasn't fair. She had raised three kids of her own. Anyway she was the one that I really relied on. Grandma was a worrier and I didn't like to worry her. She had had seven, and she was the oldest of a family of eleven.

I tried to breastfeed the babies. Everyone told me it was the best thing. That was fine as long as I lay in the hospital, but when I got home it was wintertime and I had time to feed them, but I couldn't feed them. The kid was crying all the time. And Auntie said give him Enfalac. Well, I didn't give him Enfalac – we had milk cows here, and if he didn't need a single cow's milk he could have multi cows' milk. And so I learned how to scald milk and all those things. I had a book they gave me at the hospital that told me how to do these things. It was called *Canadian Mother and Child*. I suppose it's way out of date now, but I learned how much sugar to put in. And I learned how to fix it so it suited the child's appetite. I think I did the best I could with what I had to work with.

Jay was over a year old the year that we put water in the house. So I had four by the time we got water in. They don't know anything about using an outdoor toilet. The older ones did for very short time. But we used to have a chemical one in the basement. They didn't have to go outside in the winter. They are fascinated by fire because, of course, they didn't grow up with fire. The first year Carl went to school we got gas heat. They never had to carry lanterns.

It is definitely a better life now. I don't think I would have lived to be as old as I am if I had had to chop wood and carry water all these years. And a warm car to ride to town in, and a warm house. You get up in the mornings, you don't have to light the fire, just turn the heat up. It makes for people living longer, it's much better. I think a ranch is the best place to raise kids. I would hate to raise my kids even in a small town. I guess it would be all right, but I'd hate to raise them in the city. My kids all still are ranching except Carl, and he works all over the place, in the country mostly because he is an electrician and he works on all these oil installations.

My husband worked with 4-H a lot, and I stayed home and did the work while he was away. Then he was president and on the board of the Hereford Association for years. When he wasn't involved with that anymore, he helped start the Junior Hereford Association, so he was on the road a lot with that. And of course I put in my two bits worth. I could remember all the motions that were made at the Alberta Hereford Association meetings and all the things that were decided because I had gone to all these meetings and just listened. I couldn't go to the women's things that they had, whether I wanted to go or not – I really didn't care that much, but still, there were some things I would have liked to go to just to be with women because I was always with men at home. But no, I had to sit by Stuart and make the notes and I had to remember everything. He wanted to sit there and enjoy himself and have some smart cracks to make to the others. Then when we'd go home, I had to go over it all with him again. So I guess I was involved. I even ran for office once but I didn't make it, and I knew I wouldn't, but I did it for one reason – because there would be more women coming along more capable than I, and they wouldn't maybe make the attempt unless someone did first.

Of course, it is important to have women on these boards. Who keeps the books in the majority of these farm families? And in some families the women plan the whole deal, or a lot of it. And who's out there helping pull the calves? All that goes together. It isn't a matter of EPDs [Expected Progency Difference] and numbers written in – it has to be practical experience. And the people who are doing those things should have a say in the rules that are made concerning the registered purebred business. Some of these rules are a little bit stupid because the right people are not running for office. The right people are the people who are too busy at home making a success of it to help everybody else make a strong association.

I was pretty much in charge of the Hereford end of things, a lot of it. Stuart bought the tractor he wanted, but he didn't always buy the bull I wanted. Sometimes he thought that he should have followed my advice, but before we went to Calgary I had already decided which cattle I was going to pay the most attention to. And he expected me to do that. I didn't go to as many sales and things as he did in the early years he would come home and tell me about it, but he could only tell me part of it. The pedigree business he left entirely up to me. Al has had considerable success lately and people sometimes give me credit, but he has been very good with the cattle. I used to know them all, but I don't anymore. I would maybe recognize the pedigrees, though.

The role of ranch women today only differs from what I did as a young wife as the people differ. There are the ones like me, more involved with the cattle and the horses and the crops. And then some of the younger women work off the ranch and send the kids to babysitters. If that's what they have to do, if that's what they enjoy doing, I think they should do it. I always wanted to have another job besides being so busy. I always wanted to go back to school. And for nine years, I could have gone back, when the first three had gone to school and before the next two. But I was always so busy – I was driving tractors, I was chasing cows. Most of the cow work is in the summer. And I was getting to go to field days a little bit, and I was going to the Calgary Bull Sale. And I got to go to some dispersal sales – I couldn't have done those things if I had been teaching school. By then I couldn't have gone back t eaching anyway because I had too many cows, too many hired men. I had ten people at the table year-round. And my husband was very strict about meals. He didn't help with them, but the men had to have a big breakfast, they had to have dessert at every meal. The girls don't do that now – they set you down to sandwiches. On Sunday everybody, hired men and everybody, made their own sandwiches. For lunch I put the fixings on the table and they made their own, but not if they were in the field.

I like to think that I've been a good rancher. I enjoyed the animals and I had a good memory – I didn't have to have EPDs. When you know how important it is that these things pay, then you're going to learn how to do the right things. You won't ever learn all of it, nobody ever learns all of it in one lifetime. I just wish with all the knowledge I have accumulated and all the things I've forgotten, that I could be working at it now in an active way – I could really raise a barn burner I think.

I've been really lucky with my health. I have arthritis is all. I can remember when I was hand breeding and had to be lifted on the horse. But I think I could manage on my own now. I was a lot worse then than I am now. I think because I get more rest now. Health problems have never really affected my ability to do what I want to do. And now that I'm getting older, I think I'll stay here until the old house falls down around me. It will probably last as long as I do. I like living alone. I don't have to cook for anybody, I eat when I'm hungry, I drink when I'm dry, and I don't have to work anymore.

JUDY FENTON

I was raised with five boys and mostly in a man's world, so I think I became fairly strong in my beliefs, early.

Judy grew up with five brothers on the Rutledge family ranch twenty miles southeast of Wainwright, Alberta, the daughter of the late Margaret and Gordon Rutledge. She has been married to Henry Fenton for thirty-two years, and for thirteen of those years, they were in the purebred Hereford business with Henry's parents and brothers. They have two sons, one daughter, and two grandchildren and are currently operating a commercial ranch at Irma, Alberta. Judy is the daughter-in-law of Doris Fenton.

MY GRANDFATHER AND GRANDMOTHER came out from Ireland and landed in Canada on the day the Titanic sank. They were booked on the Titanic but it was overbooked, so they came ahead on a different ship. They were not married at the time but they both travelled on that ship. They were married in Red Deer and moved out to the ranch at Monitor in 1914. Grandfather and Grandmother had five children,

JUDY FENTON AT THE BATTLE RIVER CABIN

of which my father was the third. There were two sons and they worked with Grandma and Grandpa to put together our ranch, which eventually spanned three townships and was a fairly good working ranch. The cattle were wintered at the north ranch and then were trailed fifty-six miles south to summer pasture and back again in the fall. The feed was mainly put up on the north ranch. In 1946 my father married my mother, Margaret Polsen, and they had seven children. On the home ranch until this year there have been three of my brothers and my stepmother, but this year my brother Lawrence purchased the ranch from the rest of them.

Mother was really quite busy having seven children in a span of nine years. [A girl, Kathleen, died at four months of age.] And on our ranch there were 1,400 acres of irrigated meadow and about 1,800 cows. So of course Mother always had a lot of hired men, a huge garden, and six kids. There was one thing, though, that was different about our ranch from a lot of the ranches up here. From the time I was born we had running water. We had flowing wells and I never ever remember having an outhouse – we had flush toilets and that type of thing – so I guess we were pretty lucky.

With my mother it was mostly house, garden, and children at that time, but you've got to understand, that was a very important role in running the ranch. We had anywhere from eight to twelve hired men at any one time, and our kitchen table was eight feet by four feet and it was filled twice at a meal. Mother had help at times when we could get help. There was help in the summer a lot of times, but not very often in the winter. And then Mother was diagnosed with multiple sclerosis when my youngest brother was a year old, and we did have help at times. But it ended up that some of us kids did a lot of housework, whether we wanted to or not.

By the time I was ten, I basically took over for my mother. I loved the horses and I had been riding since I was two, so

I spent as much time as I could outside with the horses. But there were things we had to do in the house, too. Even when Mother was well, Dad always put breakfast on the table for all the men and all of us at about five in the morning. When mother was not well, we did the dishes after breakfast, threw a roast or something in the oven for lunch, peeled potatoes, and got out the door to school. We came home to find a kitchen table full of dirty dishes to wash, and then we made supper. There were times we had help, mostly in the summertime we had help because there were always kids from school who wanted jobs and so on. At fourteen years of age I had a dishwasher. My aunt had come up from the States, and she said to my Dad. "Why don't you have a dishwasher? You have all this water and they have machines to do these things now." Father got on the phone and had a dishwasher installed. Of course it was just big enough to take all the plates and cutlery, but that was a big help. They all thought I was pretty spoiled.

Some of my brothers helped in the house, and actually they all did different things and they were very good at it too. My oldest brother can make bread like you wouldn't believe. He wouldn't admit to it today. When Mother was still home, they all participated. After Mother became hospitalised, it mostly fell on Ken and my shoulders. He was the youngest and he got stuck in the house with me, whether he wanted to be or not. But he's one super cook and a very good dad.

When Mother went into the hospital to stay – she went into Edmonton originally for physio and different things – and she would come home maybe two weeks of the year, and that was probably when I was about thirteen. By the time I was twenty-one, my mother was gone. She was forty-three years old. It was difficult. And it's interesting, because I can remember Mom healthy but my brothers can't. The three oldest of us remember a totally different mother than the three youngest. They always remembered her ill – I don't think Ken can remember her out of the wheelchair. Ken was twelve and Lawrence was thirteen when Mom died.

I had a tendency to be more like my father. The cattle always have intrigued me and I've ridden since I was two years of age, so the horses were always very important to me. Any spare minutes we had, they were spent on horseback, so I was really pulled in that direction. When I went to college I went into animal health. Then I was fortunate enough when I married, I married into a family where my mother-in-law was very much in charge of what was happening with the cattle, so that helped a lot.

I think my dad was very proud that I had gone in the direction I have. My father was never of the inclination that a woman couldn't do anything she wanted to do. That was kind of amazing in his time as well because there were a lot of women who men thought just couldn't do these kinds of things. And my father thought that I could do anything if I'd just put my mind to doing it.

My grandfather often laughed and said that when he came to Canada, he came with the smell of three universities on him. So Father thought education was important for us. He was self-educated – he only completed grade four and from then on he decided he was on his own, and he read a lot. He was very well read and enjoyed books immensely. He thought you should be educated, but he never thought that farming and ranching was an operation you had to get away from. I guess he instilled in us that everything he did on the ranch was a hobby – he enjoyed everything he did. He kept bees, he gardened, he was kind of an all-round type of person.

We went directly into Wainwright to school, but the ranch was south and east of the Ribstone Creek, so therefore the buses couldn't come in to us. So we were two and a half miles out to the bus to begin with. By the time I hit high school, Dad had got it arranged so the road was built in far enough that we were only one mile to the bus. By the time I hit grade eleven, Dad had convinced them to build the road one more mile right to the yard.

In high school I was missing somewhere up to thirty days out of the year to stay home and help on the ranch. In Centennial year, our Centennial project was to take steers to the Toronto Royal Winter Fair. And I decided under no circumstances were those steers leaving home without me. For once, I got my way. I missed quite a few days of school, but as long as I could keep my marks high, I didn't think it was a problem. But I knew it would be a problem if I had announced to the school before I went that I was leaving to go to the Toronto Royal. I thought it was an opportunity of a lifetime, so away I went. When I returned, I was called into the office immediately, and the principal said, "Rutledge, you realize you have missed twenty-two days this year and I can take away all your credits?" I said, "Sir, you can't take my credits as long as my marks stay where they are, and furthermore, I've probably learned more in twelve days at the Toronto Royal Winter Fair than I have learned in twelve years at this school." It was not very much appreciated, but it was accepted. So sometimes it's better to ask for forgiveness than to ask for permission.

From high school I went on to Vermilion College and took an animal health course, and when I completed that, I was married that summer. That was just one year, so I was nineteen when I was married.

College was quite an education for me, as I had never lived with a bunch of women. It was a segregated dorm, the women in one and the men in another. I had never lived in that kind of an environment at all, and it was quite a shock to my system. It was probably the biggest education I got. There were things about it that I liked but there was a lot about it that I did not like. The nitpicking and that type of thing just didn't go over real big with me.

I was raised with five boys and mostly in a man's world, so I think I became fairly strong in my beliefs, early. I don't think life would have been any easier if I had been a male. In some circumstances it might be all right, but for the most part I guess the only thing was, being a female and being a rancher and being a wife and a mother as well, there are a lot of extra little things that you have to get done, or that you're expected to do. But it's a different society nowadays – the men are taking more of a role in the house and in raising the kids than they did when we were first starting our family.

When I married into the Fenton family, they were a purebred Hereford operation and they had a small commercial herd as well. My mother-in-law was pretty much in charge of the cattle and she pretty much chose the direction of the breeding cattle. She and Stuart [father-in-law] and Henry [husband] chose the bulls together. There were hired man to look after and that type of thing. My mother-in-law didn't like to cook, but she did it when she had to and she was a very good cook, too. She had a different role than a lot of women in the industry, she was a lady before her time, and she was very good to learn from. She took really good care of her grass and she took very good care of her cattle, and she kept on top of pretty much everthing that was going on. Her record keeping was phenomenal and her honesty was way above what anybody could imagine. If a calf was born December 31st, it was registered as December 31st, not January 1st of the next year. We had an old Scotsman who worked for us and he would try his damnedest not to let her know if something popped out a few hours before it should have, but somehow she had a way of telling, so he didn't get by with much.

Before I was married, I probably did very little in the way of putting up feed. I was more into looking after the house. When it came to cattle, there was the same type of thing. I had to help move cattle – I rode and I helped at branding and so on. Then here I got a bit of an education and I became able to bale and all kinds of things I should never have learned. We put up silage, and so it kind of became a system where wherever you were needed, you stepped in and helped.

At that time we were putting on one sale a year, and then it became two sales a year and it became a bit of a treadmill. We started in January with calving, in March had a bull sale, in April turning bulls out, the end of April into seeding, then haying, then silage and combining, and back into a fall sale in November. So there was a lot to do and not much time to do it in, and it became apparent that our kids were growing up before our eyes and we were not spending very much time with them. So we decided we would be better to go on our own and go into commercial cattle, and it did free us up to appreciate what we were doing and enjoy our children. And we enjoy it now.

We've had a program where we have bred a lot of heifers summer after summer. We did that for about sixteen years. The one year, we had 1,300 head in four different places, and every day we knew we were riding. A typical day in the summertime – we would saddle up, there'd be horse work to be done, cattle to be checked, we tried to check everything every three days. We are a bit spread out and there's always things to be treated, something needs to be done. We only farm enough to feed the stock we have, we sell no grain and no feed. All our income comes strictly off the cattle industry. So pretty much it's checking, treating, or whatever. And we come in at night and make supper. But that's got much better too because Jordan is a super cook. I did train my children differently than their father was trained. I don't think my husband ever really had to make a meal. I guess he could cook a steak or open a can of beans – he wouldn't starve. That was when the kids were all teenagers that we were running a lot of heifers. It made for good horses and good kids because they always had a great summer job. I don't mind cooking, I am not great on cleaning. I think part of it is that I would like to have had someone to train me a little better how to do things easier. I know there must be better ways. I think I missed out there somewhere. The other thing, down home there was always so much ahead of me that I didn't know where to start – it was just overwhelming. And it gets that way here sometimes.

We have three children. The oldest, Gordon, is thirty, the second one, Jordan, is twenty-eight, and Jennifer is twenty-six. And we have one grandchild, Miss Sarah Jo, and she is a full year old, that's Jennifer's daughter. [Jennifer now has a boy, Jack, as well.] For some reason or another she has really taken a shine to Henry, and whenever she sees him, her little arms go straight up to him. It's quite interesting because Jennifer says she doesn't remember her dad being like that with them. I just tell her that it wasn't that he didn't love them, he was just too busy making a living. And now he can enjoy the grandchildren.

After the kids were born, I did pretty much the same work as before. In the summertime when we were over at Fentons, we silaged and did those kinds of things. It was quite interesting. My father-in-law did not have much time for children and he had no patience, so it was interesting, you know – there was no way that a child could ride in the tractor with its father, but it could ride in the silage truck with a mother who had never driven one before. So usually in the summer I had a neighbour girl come over and stay with the kids because they got tired of riding in the silage truck – it's hot and dirty. Our kids rode from the time they were little, and they trailed along behind us when we were moving cattle. When we moved over here, Jennifer was seven years old, and the neighbours couldn't imagine how we could move cattle around the way we did with three little kids and the two of us. Our kids knew they could do it and it was just expected, and they just did it. The first year we kept cattle in the Army camp out here, Jennifer had an old horse and Henry and Jennifer and I went out to bring in the bulls. Jennifer had just turned eight. The Army camp is two hundred and fifty sections and no cross fences, just a perimeter fence. On our side of the river there are twenty-one sections. So we got out about a mile and a half and here was a cow with foot rot. Henry didn't really want to leave her but we had to get these bulls in, so he asked Jennifer if she could get that cow home. Jennifer told her father that yes, she could take her home, so we left her to bring that cow and calf in and we went on to gather up the bulls. We figured she'd either have that cow at the gate by the time we got back or else she'd know where it was, but she wouldn't have left it. By the time we got the bulls gathered and back to the corner gate, she was there waiting for us with the cow and calf.

I love the cattle, I love the scenery – there isn't a day go by that I can't appreciate nature. We have lots of wildlife. I guess I used to garden a lot when the kids were little, but when we went into the heifer program, Henry and I talked about it and decided it was a lot more important to look after the heifers and look after them well and keep on top of them. Henry said, "Mrs. Tschetter can grow you a garden any time you want." And that was just what I needed to hear. Our nearest Hutterite colony is about eight miles away. So after that my garden became fourteen Saskatoon bushes and I cut the grass.

I was very involved in cowboy politics for a while with the Cattle Commission and BIC [Beef Information Centre], but when I stood back and took stock, it was over thirty days a year between the BIC and the Alberta Cattle Commission and Canadian Cattlemen's Association. That was just meeting days, thirty meeting days a year, and they were all over Canada. I enjoyed it and I liked to know what was going on. I believe that the men on the Alberta Cattle Commission and on the Canadian Cattlemen's Association were equally respectful of any input that I had to contribute as they were to any other member's input, man or woman. But it really began to pull and I just couldn't do it year after year. I just decided I had to let go of it. There are parts of it I miss. I met some very good people and I certainly knew what was happening in the industry.

Most of my friends are cattle people. Both men and women. I've kind of always lived in a man's world. It was very interesting. I've worked on very many committees and I have found that working on a mixed committee of both men and women is much easier. I just find that when you sit in a meeting with men, you can sit across the table, you can disagree, you can argue your point, then you can win, lose or draw on the vote, then you can walk out of the room and have coffee and be a friend. But for the most part on a committee of women, you can sit across the table and you argue your point, then if you happen to win it, you may not be able to go have coffee and be friends afterwards. They take it personally – that's the problem, they take it personally.

I guess for some women working off the ranch is a bit of a release, but I think our children lose out. I really do think that. They talk about us as "baby boomers," but I think in years to come they will talk about our kids as the "day care generation." They talk about how we need more money for daycares. No we don't. We need more money for some of these moms to be able to stay home and look after these babies. Those years before they hit school are the most important – even if moms stay home until the kids go to school, the basis is formed. You think about it. There's a lot of these kids where Mom goes out to work, the baby goes to day care, they come home at night, everybody's tired, they don't want to discipline these kids because they haven't seen them all day, and little Johnny needs this and this, and they feel guilty. Therefore the kid loses out. They're finding that in the school systems and it is sad. I think our priorities are really screwed up – I think it's really hard on the kids. Some mothers just want to work and they don't want to stay home, and I can understand that, too. But they don't need to stay home seven days of the week. Even if they are home just the majority of the time when the child is awake. Mom needs a break once in a while, but I don't think they should be fostered off. When you think a woman can have a baby and in six weeks it is in day care, it's scary. How is that child going to relate to people, and how much do mom and dad really miss out on?

The day before Gordon was born I was feeding green feed bales. When Jordan was born we were just finishing harvest and when Jennifer was born, she was born September 10th, and the day she came home from the hospital we were out in the field with the combines. With Jennifer I had a little trouble – my blood was low so I had these little fainting spells once in a while and that didn't go over very big. We were branding up north, probably in May or June, I was in the corral and I don't know whether it was the heat of the irons or something, but all a sudden I knew I had to get out of that corral. And I crawled out of the corral, then I flaked over, and I had some people pretty upset with me. That summer I spent in the air-conditioned car. I went for parts; I ran all the errands. That is the only time I can remember my father-in-law babysitting. It was the July long weekend and we had hay down that was right ready to bale, and Jay and Al [younger brothers-in-law] were playing ball. Stuart wanted to go to the ball games, this hay was ready to go – no ifs, ands, or buts – but Stuart really wanted to go to the ball game. So I said that I would bale for him in the air-conditioned tractor, but I told him he would have to take my boys to the ball game because I said there was no way I would sit in that tractor with two little kids for the full day and me seven months pregnant. That day we put up a full quarter of hay, I baled it and Henry hauled and stacked it, and then we had rain for two weeks. The boys still remember that day. They said every time they came near their grandfather, he handed them another two bits to go buy something so that he could enjoy the ball game without two little kids bugging him.

I think the ranch is a great place to raise children. I think you have to be conscious of safety and that type of thing. Our kids had a few wrecks along the way, and we were fortunate they weren't serious. But I guess that can happen anywhere. When I was a teenager, one of our hired men was involved in a rape in Wainwright. He didn't last on our place very long, but it was a real scare. I went to school with the victim, so it was pretty bad. I think the women and the kids were fine with her, but her dad was awful. Her dad was absolutely awful to her. But he was not a very nice man to begin with. It affected her for a long time. She no longer lives around here and I think that's a part of it. She would have been fourteen at the time. There was a stigma attached to it.

I don't know whether my kids have been better off than I was or not. Because we were a large family, we pretty much made our own fun, and any spare time we had we had a lot of fun. Because we had a lot of horses, we had a lot of kids come out from town. On weekends we had extra kids all the time. We probably had more time just to be ourselves and play. Today's kids sometimes are so over-organized that they don't have time to sit and think or do anything for themselves, and then if they do, a lot of them are sitting in front of a computer or a video game or something like that. Our kids, of course, didn't have the opportunity to do those kinds of things, but they knew they were expected to do a lot of things here, and they were also given the opportunity to travel and do other things as well.

Nowadays you're not allowed to spank, but of course, our children got spanked. They did not get beaten, but they got spanked and they learned when things were dangerous, so if that's what physical abuse is, I guess they had their share of it. I guess emotionally, at times, I guess we all were. In times of stress, you're in the wrong place at the wrong time and tongues get a bit sharp. But whether it hurt them or whether it was just a learning process? There are times probably when I grabbed my kids and gave them a smack on the seat of the pants more from fear, and to let them know that they were in the wrong place at the wrong time. I can remember with Jennifer when she was just about two years old, and I looked out and she was in a pen with about five hundred horned heifers, and they were all around her. I loaded her in the vehicle and I went to town and I got a harness and I got some clothesline cord and I tied her up. She was given a range of just so long and she could get as far as Bob's trailer. Stuart and Henry were both furious with me. They told me she was not a dog, and she wasn't an animal, and blah blah blah. But I just told them, "I want a live child, I don't want a dead one." It took her about two days to become range broke. And after that she knew exactly where she could go and where she couldn't go. If that's abuse, I guess she got a little of it. But I don't think any child should be physically or emotionally abused. I don't think berating them does much good.

Now, with the socialization and the computer age, things are a lot different. Not better or worse, just different. Some of it is a lot better, like I can throw clothes in the washer and dishes in the dishwasher, walk out the door and come back in and it's all done. I don't have to sit here and do it. I guess we have an awful lot more technology and modern conveniences today than in my grandmother's time.

Ranching is a nice lifestyle – it's hard to start out though, if you don't have someone to help you get going. I don't see anything wrong with ranching as long as you want to be a rancher. If you want to be a doctor or lawyer or whatever, that's fine too, if the opportunity is there. I would like my grandchildren to have the opportunity to make a choice. I guess we're building here for whoever wants it in the end. This ranch we kind of put together, and at times Henry thinks he would like to sell it and move on. He would like to get a place that is all in one piece, that the kids could be involved in. Right now there's not enough here to support them all, and it's kind of testy to pull it all together. But it works quite well, until we get too old to trail the cows.

Jordan is here on weekends – he is teaching full-time but he is here on weekends helping. I think he would like to ranch too, but it's getting to be such a tough racket that you have to have something to fall back on. That's why we wanted them each to have some kind of education. I guess the other thing is the freedom. If you have an education you can go do something else, you don't have to stay on the ranch. And if you stay here, it's because you want to be here, you are not trapped.

I guess to be successful at ranching, you really have to like the land and like the lifestyle. When I look around me here, Henry and I are one of the very few couples in our immediate community that don't have one or the other working off the farm. Whether that's because they want more or whether it's just to hold the place together, I don't know. I guess you have to be able to roll with the punches, too. Right now the cattle industry is booming, but we have seen it bust and it will again. You must be willing to get your hands dirty and do whatever there is to do. And sometimes stick your nose in. I think it's more successful if both man and wife are very involved in it and interested in what is happening. Always, two heads are better than one, then you can work off one another's strengths and it helps.

I don't expect to have a lot less involvement in this place over the next few years. If the kids become involved, then probably I will have to step back and let them be involved. I think that's important. I think they have to take a hand in it. As far as the physical work goes, I hope to be like my mother-in-law and still ride at eighty-seven and keep an eye on what everyone's doing and tell them when they're not doing it right. I was pretty fortunate when I got Doris, I think.

I could probably retire when the time comes, but I'm not sure that Henry can. I have a lot of interests in a lot of things – there's a lot of things that intrigue me and I don't think I'll live long enough to do all of the things I want to do. But Henry really likes his cows and he loves his horses and I don't think he would ever fit in town.

A year ago in January I was told I needed a hysterectomy. I had had a bad wreck with my horse about four years ago and it had really torn things apart inside of me, and I had a lot of adhesions and everything else, so we knew we were walking into quite a mess. I had my surgery and I behaved myself for the full six weeks before calving and everything was going good. We went to the Calgary Bull Sale and I felt really good. I had no pain, which I had been dealing with for years. I got home from Calgary and we discovered that I had thyroid cancer. So it kind of set me back again. I found out on a Tuesday and on Friday I was in pre-admission and I had my surgery on Monday. Gordon came home and helped Henry, and Jennifer came home and helped me. They were putting cows and calves out to pasture, and for the last twenty years I had put cows and calves out to pasture and I had to stand and watch out the window. That was hard. I'm doing really well now – they just do a blood test once a month for six months, and I'm doing fine. We were lucky with the cancer – it had a hot spot in it and it hadn't run yet, so we were really lucky. But it's pretty spooky and it kind of threw everybody for a wrap.

Lorraine is a real strength to me, my stepmother, she's always been a real equalizer and stable, and that helps. I'm not a religious person. My mother was brought up very religiously and so was my grandmother. My father probably read the Bible more than most people and he could quote the Bible, but we didn't attend church – for the most part we were just too far out and just didn't get there.

It's hard to say right now what things will be like for ranch women in the future because the cattle industry is booming, but when you look around us the agricultural segment is not very well looked upon. We're doing things better and better all the time, we're more efficient and that whole bit. But really, as a whole we are at the bottom of the scale, whether it's politically or whatever. One time there was a lot of pride, and I think we do still have pride in what we are, but there are a lot of ranch wives who won't say what they are.

ROSE BIBBY REPAIRING CORRALS

ROSE BIBBY

Sometimes the real story comes out.

Rose Bibby and her husband, Garth, live west of Westlock, Alberta. Rose has become widely known for her cowboy poetry, which is infused with the same sense of humour that directs Rose's life. Rose's poetry is more than just the story of a ranch: it is the story of a ranching relationship and how she has coped with issues within that relationship. Rose's writing is gritty, but humour is her guide and laughter her gift.

OUR WORK NEVER STARTS on the second of January, more like the first of January – but there is sort of a cycle. We've been here now since 1970, and now it has gotten so I'm not outside as much because we're winding down and the work is becoming a little less. It's also become more mechanized, so the ranch wife doesn't have to be outside quite as much. I'm kind of the official gate opener now. We feed our cows in the winter with a tractor, so we're out there getting those chores done. We keep our calves over and have them to feed through the winter, too. The horses get a bit of a break through the winter. We calve later on in April and May, and I'm out there checking – we still check them at three in the morning because we can't stand not doing that. We calve them on forty acres, so we get to ride around or drive around, whichever, depending on the weather. So I'm checking the cattle, helping where you have to help, tagging, treating, all those sort of things.

We sow some crops. What we like to do here is to sow just enough grain crops to rotate our pastureland. The last two or three years of drought, we've ended up with way more grain land than we'd like to have, and that's just the way it has worked out. And meantime, I'm kind of the checker for which fields are dry. I usually rake the hay. Lots of time to think out on the rake. I can write poetry, sing songs. When you're on the rake, you can sing songs and you think nobody can hear you and what you're actually trying to do is sing above the noise of the tractor. During summer it's haying time, and our life is focused around haying and fencing. There is always lots of fence to build or fix. So many outside jobs seem to go smoother if there are two people working, and that includes calving, sorting, and weaning. Of course, we have our branding in between there. Branding is an exciting time because the whole family and friends get involved. It becomes kind of a social occasion, but what a wonderful way to get together and all share the work, share the food. If you get enough people at a branding, of course, there's only one person that really works hard and that's the cook.

It seems that at each ranch there's kind of something special the cooks do that maybe the other ones don't. My speciality is sour cream rhubarb pie. Sweet and sour meatballs are another specialty here. When we started out with the branding, we would have stew and potatoes and pie, always pie. But now I've sort of evolved so I am making something I can prepare the day before and have it ready so it is easier the day of the branding. I started out with stews and that sort of thing, and I ended up doing things that were my thing, and possibly that was because we had something different at our place, and we changed things. Earlier on we had big stews and fresh baked bread. There's nothing like fresh bread. Now, I don't know, it's kind of an evolution. We just changed and worked with what's handier.

The woman's other job is to make sure all the needles, disinfectants, and vaccines are ready. I'm kind of the "gofer." A lot of ranch women are the "gofers" because they know where things are and they can run for them. Maybe it's just because you always keep track of those things. That's partly because you're expected to, partly because women seem to be able to do a lot of things at once. And so when you get a day like a branding, it seems to be easier for a woman to put all these things together and make sure that they're there. Why that became her job, I am not sure. Maybe because she would do it, that's why. I'm kind of talking in circles.

Our oldest daughter never learned to rope, so there was never any talk of her working up to the roping at branding, but because she's always been involved with the cattle she can kind of choose what she wants to do, choose in the sense that she can do a lot of things. She will quite often do the branding. She likes that. This year she wanted to be at the pen gate, watching that nothing gets out. She can do any of the jobs – the needling, the branding, that sort of thing. That's Christine. Leanne, our second daughter, always liked to be riding. She can rope and tried her hand at that but never at a branding, just at other times when we were working the cattle. And she likes to brand, that's the job she prefers. The girls have always, always learned to do every job that needs to be done.

Our third daughter, Carmen, was always very tiny but she wanted to do what everyone else was doing, so she chose to have her father show her how to castrate and that is her official job. Over time you kind of get a job that you do, and when the next branding comes along, you know that the same person is still going to do that job unless there's a shortage of some people in some area and you have to take over. This year our grandsons were more help – some were wrestling calves and the eldest took a turn at roping.

The big thing is that Garth and I have gone off the horses and gone on the stage. Actually, for twelve years I had written a column in a newspaper. What happened was we had a forage association here, and Garth was involved with the association and started writing a column in the paper called "Forking and Coiling," and he wrote that speaking like an old hillbilly. Then they were looking for more material for the paper and I started submitting some of the poems I already had written for various occasions. What happened next is I began to write poetry that talked about what we were involved with here. The fact that when we get together to work cattle, it's always handier to have the neighbours come in because there's not so much swearing – he doesn't swear at the neighbour's wife. And some of these things would come out in the poetry. I would then submit these poems to the paper and get favourable comments, so I started putting down more of the events that happened on a daily basis around the ranch. Garth's quite volatile when it comes to things happening, so you always hear it in the language. So a lot of the poetry stems from how we work with each other and how he uses the language.

One year, it was calving time, I had been to town and stopped at a neighbour's who I would hardly ever visit. I had always written anonymously in this newsletter, and she had just discovered, after twelve years, that I was the Hayshaker's Wife. And she said, "I wish I would have kept all your poems." So I came home and said to Garth, "What would you think if I were to put some of these poems in a book?" So we did. Then we decided we would go down to Pincher Creek and see if what I did fit with what they did there. I had no idea.

I went to the Pincher Creek Cowboy Poetry Gathering and I was on the open stage there, and within a week I had two calls to come and present my poetry at two other events. I did that for two, two and a half years. Garth was driving me to the cowboy poetry events – a lot of my poems were about him, so he said he was my prop. I'd tell the stories and they'd laugh at him. This got to be a bit of a drag. So on the way to a show at Cranbrook, I had written some poems that I thought Garth could present, but he was not comfortable doing this. On the way back from Cranbrook, we dreamed up the idea of the "He said, She said" poetry. I, for instance, would tell how much help I was on the ranch, and he would question how much help I would be:

It started right early this morning
in fact, as we went out the door
The hurry, the cursin,' the panic
that comes where there's cattle to sort.

In fact whether treatin' or brandin'
or penning or culling's the game
Or calvin,' or countin,' or haulin'
the problem is always the same.

The verbal harassment's unending
as by now I should understand
How that old cow's brain is workin'
as well as what's on "his" mind

It's a pain in the butt, this harangin'
I get tired of hearin' it all
But I found through years of observin'
You know, sometimes he don't swear at all.

In fact when we call in the neighbours
to help with the brandin' and such
There's considerable less noise and tension
and he don't curse near so much.

So I've kind of gotten this habit
when I know the day will be long
To beg for some neighbourly sharing –
this tactic has never been wrong.

I get a break from the blue sound and colour
that seems to cloud up my life
And off his tongue the honey comes drippin'
He don't swear at the neighbour's wife!

That's one and then Garth would come back with one like "I thought she wanted to be a rancher's wife," and he would talk about that.

When I first started doing this, I was the only one on stage, and I really felt that women tend to laugh at themselves a little quicker than men, this is my observation – only my opinion – and especially if a woman is asking a man to laugh at himself. I've found that women of the generation before me don't talk about their husbands and the things that happened as much as what my generation and the younger ones will do. So one time this little old woman wanted to tell me something but she really didn't want to come out and say it publicly, so as she strolled by me she put her hand up by her mouth and just to me said, "They're all alike." I have always treasured that moment. It was something she had to say, but she couldn't say it right out. I found that very interesting and afterwards I compared her with my mother, and my aunts and my grandmother. They would have been the same way. They wouldn't have said, "Okay, this is something that happened at our place and it just makes me so mad, and I can't stand it." They wouldn't have said that. In the poetry, it comes out and it's not a direct assault on the problem. That was a bit of an eye-opener, and it kind of reinforced what we were doing but not in a derogatory way.

I do a lot of conferences where women are away for the day for some R&R. One of the reasons I think they want to go is to remove themselves from the place for the day so they can talk about the issues, the things in their lives. And one of the things I have been able to do is send people home feeling good about their jobs and what they do. That became very important for me, and I was happy that I was able to do this – part of the poetry went in that direction.

A lot of my poems were written when I was outside. It used to be my job in the winter when we fed round bales to take the strings off the bales. I wrote many a poem while traipsing around these bales. I've also written on the tractor. But my most popular poem I wrote when I came in the house one day, entirely frustrated, because all that seemed to matter here was his horses, cows, and his dog. I ended up writing a poem called "The Love of a Cowboy" where he had proposed and asked me to share his life with his horses, cattle, and dog. It naturally ended up being a love story in that sense, and so it sometimes seems the story changes as I get it on paper. Sometimes the total frustration comes out, but sometimes the real story comes out.

We found that when we were doing this – I would tell one side of the story and he would tell the other side on stage – the people in the audience react because the men are sitting there with their wives and they can relate because they have been in these situations. We have had people come up and say, "How long was it that you were standing at the end of our lane?" because so much of the stuff that we portray on stage fits. And it doesn't matter whether you've hung wallpaper with your wife in the city or you've worked with her on the ranch, it fits.

Sometimes we can do three or four events a month, private functions where we entertain. We did a wonderful event – a keynote address at a farm women's conference in Grand Prairie last fall. I talked for a while about the role of the farm wife and the people who had influenced me in my life – my mother especially and some dear friends. I talked along those lines and then we did some poetry back and forth. It's been a lot of fun. We say a lot of things a lot of people would like to be saying, but they're not saying. We can get up on stage and make a joke in a sense. Even though we're kind of poking fun at each other but not really being rude, just telling the truth in a joking manner, we have had many moments where people will say you can see the love come through.

What do I foresee? That's a tough one. When we were growing up, Garth's father, more so than mine, did not encourage Garth to have anything to do with the land. He just felt Garth should be off with an education and doing something else. And I'm not sure why that happened because when his father's health was gone and it was time for him to retire, he didn't want to leave the place either. But it becomes harder and harder to make a living on the ranch with all the other things – the costs, the politics – that seem to be at play. Thirty years ago it was hard, but it seemed that with hard work you could make it go, and it's still hard work but there's so many other things that come into play now that it makes it very, very much more stressful. And again it's hard to encourage younger people.

Our oldest daughter Christine, and her husband – in fact they have just purchased land in the north so they can have a bigger place. Ranching is what they want to do. And right now they have a fourteen-year-old boy who thinks that's what he wants to be. Like I say, it's been good to us and we've had a wonderful life here, but it becomes ever increasingly difficult to make your living from the ranch. It's hard to encourage anybody to do this.

The best part, I guess, the main thing for us, for me, is that I can get up in the morning and do – I was going to say do what I want – but I can be in the country away from the city. I never liked being in the city. I think the freedom of being out here is the best part. And the other thing is being able to have your family involved with what you're doing – that has been really good for us. We still work back and forth with our children and that has been positive as far as I'm concerned, watching the kids and teaching the kids the things you do on a ranch. I don't care whether they end up being architects or what they end up doing, all the things that they learned on the ranch – being self-reliant is a real plus, because they learn to work and be responsible here, and it will help them as they progress in whatever they decide to do.

ABOUT THE EDITORS

THE INTERVIEWERS: JOANN JONES-HOLE, SUSAN AMES VOGELAAR, THELMA POIRIER, ANNE SLADE, DORIS BIRCHAM

THELMA POIRIER

Cliff Anderson's sister

Thelma (Anderson) Poirier, the youngest of thirteen children, was born in 1940. She grew up on a ranch twelve miles south of Fir Mountain, Saskatchewan. In 1959 she married Emile Poirier, also of Fir Mountain, and they raised three sons: Perri, Gary and Robin. Perri is a computer technologist at Lafleche; Gary is a systems engineer at Calgary; and Robin ranches on the homeplace. Thelma has always been interested in history and has been active in the Wood Mountain Rodeo-Ranch Museum. She has edited four ranch books: *Beyond the Range*; *Cowgirls: 100 Years of Writing the Range*; *Grass Roots*; and *Wood Mountain Uplands*. Other published books include *Grasslands*, *Rock Creek*, and *The Bead Pot*. Thelma is a planner and an organizer. Her innovative ideas flow freely on to the pages of her books, whether those books be history, non-fiction or poetry.

AFTER I WAS MARRIED I became involved in the Poirier family stock farm. Often Emile and I took our boys and made a picnic out of fencing or checking cattle. I loved the grazing lease with the cattle, the wildflowers, the burrowing owls, deer, antelope, sage hens, the hills, the creek, and the grass. After a few years of teaching, I became even more involved with the ranch, buying land and cattle and a horse of my own. Over the years, our stock farm became more of a ranch. Summers and winters we did the usual work – trailing cattle, rounding up, checking cattle, and feeding. I felt displaced when we started feeding with a round bale processor a few years ago. I lost touch with the cows until spring. Then came calving and I renewed my relationship with the cows. Birth and rebirth, that is the best of all.

After much deliberation, Emile and I moved to Glentworth, a small village north of the ranch. I wanted to stay where I was on the ranch, but that wasn't practical. Our son Robin and his family live there now. It was hard to leave the only home we ever had, the place where we had the best years of our lives, where everything represented our life's work.

Change is inevitable. Only a few things remain the same – the land and our love for it; our family including ten grandchildren and these computer keys. I continue writing; never sure what story will come next, what poem will surprise me.

DORIS BIRCHAM

Doris (Shearwood) Bircham was born in Maple Creek, Saskatchewan in 1938. Growing up in the Cypress Hills gave her a love of the land and the outdoor life. After becoming a nurse, Doris worked for a year at the Maple Creek Hospital. Following her marriage to Ralph Bircham in 1959 and a year spent in Edmonton, the couple bought the ranch where they currently live. Their son Wayne, and daughter Dena are both married and ranching with their families in the area.

Doris was co-organizer of the Maple Creek Cowboy Poetry Gathering for twelve years. She also taught First Aid, CPR and worked with Saskatchewan Home Care. In 2003 she received a volunteer award from SaskCulture. Doris's kindness, empathy, and good humour make her an especially valued friend.

THAT YEAR WE WERE in Edmonton, we spent every weekend off looking for a place. Ralph had five thousand dollars saved, fifteen cows, and a new truck. I had my RN, some petty cash, and a second-hand piano. We purchased the Newkirk Ranch south of Piapot, Saskatchewan in 1960.

We lived in a three-room house built out of scraps from the 76 Ranch shacks. It was stuck in the side of a hill with a basement underneath. The mice ran in and out. We had a wood cook stove and heater, so in winter we ran in with wood and out with ashes. Our kitchen was twenty-four feet long to the bay window and the floor sloped half an inch to the foot. On December 30, 1966, we ordered a new house package. There was no rain in '67 so there was no crop or money, but there was time to build.

We milked cows because that put groceries on the table. Chickens – I hated chickens so didn't have them any longer than I could help. I raked hay with a team and we fed the cattle with a team. Once when Wayne was a baby, we cut a field of oats with the binder, made him a little nest on the floor of the truck cab and checked him every round.

We trail cattle to the hills and back again, spring and fall. Before Dena was born, I remember barely fitting behind the steering wheel, following behind the cows, trying to manoeuvre a two-wheel-drive truck through sticky gumbo in Dead Steer Coulee.

Both our children were in 4-H and both started building their cow herds at home. Now we have four grandchildren who live close enough to regularly enrich our lives. Ralph claims he's retired, but he's still actively involved and I still enjoy heading down to the corral to help out. I've always loved to ride but have had a few wrecks along the way. I celebrated the new millennium by getting bucked off in the spring, and my horse went down in a sinkhole that same year. Although it was scary, we both came out uninjured.

Poetry is something I've dabbled in since childhood. Cowboy poetry is fun, and I especially like working with school students. Most important are the friendships made along the trail. My writing has appeared in a variety of magazines, anthologies, and two textbooks.

As ranchers, we're governed by markets and weather. Stress levels seem to be increasing and I worry about that. Where I live, I'm surrounded by remarkable, innovative ranch women, and I truly regret that this book didn't have enough pages for all of them.

JOANN JONES-HOLE

Genuine warmth, down-home hospitality, a commitment to the job at hand, and a wacky sense of humour: these phrases could all describe JoAnn Jones-Hole. JoAnn (Hendricks) was born in Calgary in the early forties and raised on a farm at Irricana, with three sisters and one brother. She was her brother's little helper. She rode, she barrel raced, she fenced, and she chased cows. Following graduation from the University of Calgary, JoAnn taught at Banff Trail School in the city. She married Douglas Jones in 1962 and left teaching to raise their family. While she was a teenager in high school, JoAnn's father died, and in 1978, her husband, at age forty, was killed in a tractor accident. When asked how she has managed to overcome tragic events in her life, JoAnn answered, "I do a Scarlet O'Hara – I'll think about that tomorrow. You just do for today what you have to do." Her writing has appeared in *Tapestry* (1996) and *Cowgirls: 100 Years of Writing the Range* (1997).

JoAnn is the mother of Catherine Chalack and Cheryl Morison, whose stories are included in this book.

WHEN I MARRIED DOUG I learned that cows come first. We rarely went anywhere if it wasn't to see cattle. We had a purebred operation, and if we wanted to have a Sunday with the children, we got out of there by nine o'clock in the morning.

Everything belonged to the ranch. We had nothing that was just ours. I believe every married couple needs something of their own, just theirs to work for and improve on. But a result of being in the purebred Hereford industry is that we have friends from all over the world, good friends.

I love children, have always loved children. My two daughters have married local boys, and my son, Brad, has just completed his master's in software engineering. Now I have five grandchildren and I've had the privilege of getting to know each of them really well.

Over the years I've been involved with 4-H, home and school, the Chinook Hereford Belles, the Alberta Writers Guild, the Beef Information Centre, and the Stockmen's Memorial Foundation. When my book, *Calgary Bull Sale 1901–2000*, was auctioned, the proceeds were used to purchase two beds for burn victims at the Children's Hospital, where I did volunteer work.

In 1980 I married Jim Hole. My husbands, both of them, have been my best friends. Both ranched. Ranching is a way of life that I want my children and grandchildren to have – the neighbourliness, the ties to the land that keep me grounded no matter what I do. We have to do our best and have faith in what we do, and not ever expect we're going to make a fortune ranching. If we make a living and enjoy what we're doing to make that living, that's great.

I count among my friends and acquaintances some interesting, admirable women who have dedicated their lives to ranching, their communities, and their families. I would have liked to interview and include them all in this book, but it just wasn't feasible. I interviewed a sampling of women who I thought would represent ranch women of today, as well as some who are individually different.

Brad Jones married Nicole Allarie in the summer of 2004.

ANNE SLADE

Some people come up with great ideas; others turn those ideas into reality. Anne (Shea) Slade does both. She radiates warmth. Anne's ability to seek out positives has helped her to deal with challenging medical situations in her family, as well as assisting her with her journey through grief following the death of her eldest son, David. In addition to being the mother of four boys, Anne has been a volunteer on a variety of boards and served ten years as a municipal councillor. She co-wrote a play about the town of Tompkins and acted as co-editor for the town's history book. She is a singer and songwriter, and has been actively involved with cowboy poetry as a poet and organizer. She has authored *Denim, Felt & Leather* and has co-authored *Pastures, Ponies & Pals* and *Pocket Poems for Kids*. Her work has appeared in a variety of magazines and anthologies and on a video. Ken Slade, Anne and Robert's second son, and his wife, Wendy, are teachers. Art Slade and his wife, Brenda Baker, are writers. Art received the Governor General's Award for Children's Literature in 2001. Brett has an engineering degree and is a rancher.

I WAS NINETEEN when I got married in 1962 and came to the east block of the Cypress Hills. The ranch was fifteen miles from the nearest small town and there were five gates to open on the prairie trail going into the ranch. Delores Noreen [see her story], who had worked for the Slade ranch for years, turned the kitchen over to me. That first summer it was my job to cook for four or five hired men. Duff, Delores's husband, taught me how to make potato salad – Delores showed me how to make meat loaf and mix bread.

At first I was really lonely, but after two or three years you couldn't have dragged me off the place.

David was born nine months after we got married, so it wasn't until one summer when we were shorthanded that I learned to run the tractor. I did all the baling that summer, and later learned to drive grain trucks, check heifers, and help with chores when needed.

I took correspondence classes and attended community college courses through the years, then completed my Bachelor of Arts ten years ago, at the University of Saskatchewan. For eight years I worked as a teacher assistant, and I still substitute occasionally. I have also travelled throughout Saskatchewan presenting poetry to schoolchildren and teaching them about our western heritage.

A few years ago I started quilting and have found it a lot of fun. We also have an interdenominational choir in our community. It doesn't matter what church it is – we're there to be supportive for special services. And I now have grandchildren, three boys and a girl, to add an exciting dimension to my life.

Ranching is such a wonderful lifestyle. The animals and the closeness to nature are important, but it's the people involved that make it such a rich life. I feel lucky to be a part of it.

Brett Slade married Karla Huber in 2004, and both are presently attending the University of Saskatchewan, where Karla is completing her degree in veterinary medicine and Brett is taking his master's in business administration.

SUSAN AMES VOGELAAR

Susan Ames Vogelaar is an instructor at Lethbridge Community College and a cattle producer east of Pincher Creek, Alberta. Like her war-bride mother, Susan embraces challenges, challenges as varied as bundling up preschoolers to help with winter chores, caring for her critically ill daughter, teaching school, volunteering on boards, organizing the cowboy poetry gathering for seven years, coaching basketball, and completing her MA after her family had grown. She is now adjusting to her teaching and part-time ranching lifestyle. Probably her most difficult challenge was finding her place in a traditionally patriarchal Dutch family, in which the brothers formed a corporation. However, as Susan says – with her candid good humour – "I'm a rebel and never could conform to the patriarchal dictates of the men."

Susan's creative spirit is revealed in her writing, watercolours and sketches. In 1997 she edited the Alberta Cowboy Poetry anniversary anthology, *Bards in the Saddle*. She has written for Legacy Magazine and her cowboy poetry appears in *Cowboy Christmas* and *Northern Range*. She has published two books of her cowboy poetry.

Susan and Jake's three children were treated equally, learning skills inside and outside of the home. Grant, who has his Bachelor of Science, his wife, Lori, and their sons, Matthew and Kyle, raise cattle. Greg, who is married to Kellie Ann, has a Bachelor of Science degree, and is a paramedic in Calgary. Angie, who is married to Tyson Mackin, is an animal health technician, and is back at university, studying for her Bachelor of Science. Susan and Jake always stressed the importance of education and the bonds with the rural lifestyle.

FOR MANY YEARS women's voices were not accepted in our society. On the ranches and farms and in the rural communities, women play a pivotal role. Often although they're the strength, their voices are not always heard, or are not accepted as equal. I've always felt uncomfortable with that, and I think that's my father's fault. He taught me to be independent and outspoken.

If we had operated as an independent family farm, I would never have worked outside the home and our children would probably still be involved in our own place. But – I think – because of the way it was set up, this corporate family patriarchy, I felt that I had to pursue my own interests.

My memories of the early years include the old shack we first lived in, where we used the front porch for the fridge; cold, that froze the propane tank in the middle of the night; and pigs, digging up our precious garden, which provided us with our winter vegetables; and calving in the middle of the night. Our children loved the rural life and community, including neighbourhood broomball parties, hockey games on the creek, and hayrides at Christmas. It gave them a sense of place and responsibility, and skills such as caring for the animals, working with equipment and understanding the concept that rewards come with hard work.

I'm looking at retiring from teaching in a couple of years. We plan to keep the horses and maybe a few cows because I can't imagine separating from them. I just hope that there will be a future in agriculture for young people.

APPENDIX

This appendix outlines the interviews done by each interviewer.

DORIS BIRCHAM

Kim Taylor	January 7, 2003
Heidi Beierbach	July 2, 2003
Pansy White-Brekhus	July 30, 2001
Mary Guenther	August 17, 2002
Ann Saville	August 14, 2002
Mary Jane Saville	February 26, 2001
Dena Weiss	June 23, 2003
Delores Noreen	November 6, 2002
Hilda Krohn	February 7, 2001
Marilyn Ramsay	March 29, 2003
Robin Ramsay	March 29, 2003

JOANN JONES-HOLE

Catherine Chalack	January 4, 2003
Irene Edge	October 5, 2000
Beryl Sibbald	October 5, 2000
Cheryl Nixdorff	October 19, 2000
Ruth Hunt	October 18, 2001
Cheryl Morison	January 8, 2003
Vernice Wearmouth	June 27, 2001
Edith Wearmouth	June 21, 2001
Leta Wise	October 8, 2001
Eileen McElroy Clayton	January 15, 2003
Pat Kerr	August 31, 2000
Doris Fenton	June 12, 2001
Judy Fenton	June 12, 2001

THELMA POIRIER

Heather McCuaig	December 2, 2002
Marjorie Linthicum	November 13, 2002
Louise Popescul	September 11, 2003
Linda Froshaug	September 12, 2002
Tammy Burgess	March 21, 2001
Ruth Pritchard	November 16, 2002
Sherri Grant	September 10, 2003
Sandy Hordenchuk	September 4, 2003
Jill Mastad	January 5, 2001
Marilyn Jahnke	June 30, 2003
Rose Bibby	June 29, 2003

ANNE SLADE

Joan Lawrence	October 21, 2002
Heather Beierbach	September 5, 2003
Christa Lawrence	October 21, 2002
Erin Bircham	April 6, 2003
Lou Forsaith	April 20, 2002
Roberta Wolfater	February 17, 2003
Lyn Sauder	October 16, 2002
Rajanne Wills	February 21, 2003
Carley Cooper	March 6, 2001
Gayle Kozroski	April 3, 2003
Clare Kozroski	February 17, 2003
Diane Catley	December 1, 2002

SUSAN AMES VOGELAAR

Helen Cyr	July 19, 2002
Alice Streeter	June 30, 2003
Doris Burton	permission October 16, 2003
Virginia Delinte	July 16, 2003
Anne Stevick	July 10, 2002

GLOSSARY

Agribition: a large international livestock show and sale held each fall in Regina, Saskatchewan

A.I.: artificial insemination

Barrel racing: a women's rodeo event involving racing a horse around three barrels

Binding: done by a machine called a binder, which cuts the ripened crop and ties it in bundles in preparation for threshing, an early-twentieth-century method for harvesting grain crops

Bloomers: full loose trousers gathered at the knee, worn by women in the early twentieth century

Bonspiel: a curling tournament

Boundary Commission Wagon Train: a wagon trek from the Red River in Manitoba to the Alberta foothills

Brandings: gatherings of neighbours to help ranchers brand calves for identification

Breakaway roping: roping in which a roper releases the rope once an animal is caught or the rope itself has a honda (noose) that releases

Breaking: newly ploughed native sod

B.S.E.: bovine spongiform encephalopathy, or mad cow disease

Bum lambs: orphan lambs

Calving season: season when calves are born

Canners: usually refers to old cows, or horses that are sold
Cat: Caterpillar, a make of tractor that runs on tracks instead of wheels
CCA: Canadian Cattlemen's Association
Coulee: a shallow ravine
Cowie: a horse that has cow sense, handles cattle well
Cull cow: a cow that is culled out of a herd
Cutter: horse-drawn sleigh
Cutting: castrating
CWB: Canadian Wheat Board
Doc o' Lena: the name of a famous quarter horse stallion
Dogie: a calf
Draft horses: work horses such as Percherons, Belgians, or Clydesdales
El Camino: the brand name of a small car with a truck back
EPD: expected progeny difference, a record-keeping tool used mainly in purebred cattle herds to predict the future performance of an animal in the areas of growth, milk, and fertility.
Foot rot: a bacterial infection in the hooves of cattle that makes them quite lame but is easily cured
Futures: marketing grain and cattle by gambling on the price in the future and locking in at that price
Gofer: a person who goes for supplies or runs errands, to go for
Greenfeed: usually a combination of oats and barley, cut and baled for feed while green, as opposed to leaving it to ripen for grain
Hand breeding: the practice of bringing cows one at a time to a bull, as opposed to pasture breeding, where the bull stays with the herd, used when there are too many cows per bull or for young bulls to save their energy for growth
Harper Ranch: former privately owned deeded land, now the property of the Kamloops Indian Band, the home ranch of Thadeus and Jerome Harper, early B.C. ranchers who owned about four million acres of land in the Crown Colony of British Columbia in the late 1800's. The name "Harper

Ranch" has been retained by the former owners, the Kerr family.

Heading a calf: roping a calf's head or horns

Heeling a calf: roping a calf's back legs

Heifer: a female bovine prior to having her first calf

Heinz 57: a dog or other animal of mixed breed

Hogsback: a sharp ridge between two valleys

Hudson Bay quarter: 160 acres of land in every township given to the Hudson's Bay Company as payment for Rupert's Land

Hutterites: a religious group of Russian/German heritage living on colonies and sharing "all things in common"

Jack: a male donkey, a jackass

John Deere Gator: a small four-wheeled vehicle used for farm and ranch chores

Kidding: the birthing process for goats

Light horse: a saddle horse or horse used for pulling a buggy or cutter

Line camp: a small cabin on large ranches where employees and travellers could camp or stay overnight

Made the cut: was good enough to qualify as being a part of a herd or group

Mugger: someone who wrestles an animal by grabbing it by its head and ears

M.D.: municipal district

Murraydale Rodeo: a rodeo in the Cypress Hills that has run consecutively for over ninety years

Open range: native grassland belonging to the Crown that was open for anyone on which to graze livestock

Pail bunter: a calf raised on milk from a pail instead of directly from the cow, usually the calf of a milk cow

Palliser Triangle: a triangular area of southern Alberta and Saskatchewan that was mapped out by Captain John Palliser in 1857 and that was considered unfit for cultivation

Player piano: a piano that is operated by inserting rolls of perforated paper, and manually winding it up to play by itself, much like a music box. It may also be played like a regular piano

PMU: pregnant mares' urine, used in the production of hormones for human use
Powwow: a gathering and celebration of Aboriginals
Prairie oysters: (calf fries) bovine calf testicles
Pullets: young female chickens
Pulling calves: providing assistance to cows giving birth
Putting down: killing an animal humanely to end its suffering
Ranch hand: a hired man or woman working on a ranch
Rake: a machine used to rake hay into windrows or swaths after it has been mown
Range broke: refers to livestock that know their own grazing area without a fence and do not leave it
Replacement heifers: heifers that are introduced into a herd and replace cull cows
Roundup: an event in which riders gather cattle from pastures for such purposes as branding or weaning
SAIT: Southern Alberta Institute of Technology in Calgary, Alberta
SARM: Saskatchewan Association of Rural Municipalities
SSGA: the Saskatchewan Stock Growers Association
Section: one square mile, or 640 acres of land
SETA: Society for the Ethical Treatment of Animals
76 Ranch: a large company established about 1887, consisting of ten ranches of ten thousand acres each, located between Balgonie, SK, and Langdon, AB
Shy breeder: a cow that does not easily become pregnant but is not completely infertile
Slough: a natural body of water with no creek or river flowing to or from it
Square draw: weekly curling games within a curling club
Steer: a castrated male bovine
Stone boat: a small sled hooked behind a team or tractor and used for hauling
Stook: ripened grain in a pile of eight to twelve bundles stacked upright
Subbing: substitute teaching

Swather: an implement used to cut hay or grain and lay it in a windrow

Sweep: a machine pulled with two teams of horses, used to pick up two windrows of hay at a time and move them to a wagon following behind

Tagging: the process of putting a tag in an animal for the purposes of identification, usually an ear tag

Team roping: two riders working together to "head" and "heel" an animal

Threshing outfit: an old-time crew of workers stooking sheaves, loading them on wagons and throwing them into the threshing machine

Threshing: harvesting grain by removing the seeds from the rest of the plant

Township: thirty-six contiguous sections of land

Tug chains: chains connecting a workhorse's harness to the implement being pulled

Two bits: twenty-five cents

Weaning: separating a cow from her calf permanently

Western Producer: a weekly agricultural newspaper for the prairies

Windrow: a continuous row of hay or grain laid down by a swather

Wire stretchers: an implement used for tightening wire fences

Wrestling calves: two people working together to wrestle a calf so it's lying on its side, ready to be branded, vaccinated and tagged, usually done at brandings

LEGACIES SHARED SERIES

Janice Dickin, Series Editor

ISSN 1498-2358

The Legacies Shared series preserves the many personal histories and experiences of pioneer and immigrant life that may have disappeared or have been overlooked. The purpose of this series is to create, save, and publish voices from the heartland of the continent that might otherwise be lost to the public discourse. The manuscripts may take the form of memoirs, letters, photographs, art work, recipes or maps, works of fiction or poetry, archival documents, even oral history.

Looking for Country: A Norwegian Immigrant's Alberta Memoir
Ellenor Ranghild Merriken, edited by Janice Dickin · No. 1

The Last Illusion: Letters from Dutch Immigrants in the "Land of Opportunity" 1924–1930
Edited and translated by Herman Ganzevoort · No. 2

With Heart and Soul: Calgary's Italian Community
Antonella Fanella, edited by Janice Dickin · No. 3

A History of the Edmonton City Market, 1900–2000: Urban Values and Urban Culture
Kathryn Chase Merrett · No. 4

No One Awaiting Me: Two Brothers Defy Death during the Holocaust in Romania
Joil Alpern · Copublished with Jewish Heritage Project · No. 5

Unifarm: A Story of Conflict and Change
Carrol Jaques, edited by Janice Dickin · No. 6

As I Remember Them: Childhood in Quebec and Why We Came West
Jeanne Elise Olsen, edited by G. Lorraine Ouellette and Ian Adam · No. 7

Alequiers: The History of a Homestead
Mike Schintz · No. 8

Far from Home: A Memoir of a Twentieth-Century Soldier
Jeffery Williams · No. 9

To Be a Cowboy: Oliver Christensen's Story
Barbara Holliday · No. 10

Betrayal: Agricultural Politics in the 1950s
Herbert Schulz · No. 11

An Alberta Bestiary: Animals of the Rolling Hills
by Zahava Hanan · No. 12

Memories, Dreams, Nightmares: Memoirs of a Holocaust Survivor
Jack Weiss · No. 13

The Honourable Member for Vegreville: The Memoirs and Diary of Anthony Hlynka, MP
Anthony Hlynka, translated by Oleh Gerus · No. 14

The Letters of Margaret Butcher: Missionary-Imperialism on the North Pacific Coast
Margaret Butcher, edited by Mary Ellen Kelm · No. 15

The First Dutch Settlement in Alberta: Letters from the Pioneer Years, 1903–14
edited by Donald Sinnema · No. 16

Suitable for the Wilds: Letters from Northern Alberta, 1929–31
Mary Percy Jackson, edited by Janice Dickin · No. 17

A Voice of Her Own
edited by Thelma Poirer, Doris Bircham, JoAnn Jones-Hole, Patricia Slade and Susan Vogelaar · No. 18

Missionaries among Miners, Migrants and Blackfoot
edited by Mary Eggermont-Molenaar and Paul Callens · No. 19

What's All This Got to Do with the Price of 2 x 4's?
Michael Apsey · No. 20